...ING THE LIGHTS ON

...ards Sustainable Electricity

Walt Patterson

First published by Earthscan in the UK and USA in 2007
Paperback edition first published in 2009

For a full list of publications please contact:
Earthscan
2 Park Square, Milton Park, Abingdon, Oxon OX14 4RN
711 Third Avenue, New York, NY 10017

Earthscan is an imprint of the Taylor & Francis Group, an informa business

ISBN 978-1-84407-798-4 (pbk)

Typeset by MapSet Ltd, Gateshead, UK
Cover design by Nick Shah

A catalogue record for this book is available from the British Library

Library of Congress Cataloging-in-Publication Data
Patterson, Walter C., 1936-
Keeping the lights on : towards sustainable electricity / Walt Patterson.
p. cm.
Includes bibliographical references and index.
ISBN-13: 978-1-84407-456-3 (hardback)
ISBN-10: 1-84407-456-0 (hardback)
1. Electric utilities—Government policy. 2. Energy policy. 3. Sustainable
development. I. Title.
HD9685.A2P38 2007
333.793'2—dc22

2007005933

CONTENTS

PREFACE TO PAPERBACK EDITION

When Earthscan told me they were to bring out a paperback edition of *Keeping The Lights On* I was pleased. They said it would be an opportunity to update the book, to take into account significant developments since the original edition was completed in early 2007.

I thought about developments. Scientists around the world now concur that the Earth's climate is changing even faster than previous prognoses foresaw. Human activity, particularly the way we use fuels, threatens to precipitate global catastrophe, not in some remote decade but within the lifetime of our children.

However, decision-making of every kind, including that for energy and electricity, is beset by ever more severe uncertainty. Arrogance and greed have led to worldwide financial chaos, crippling investment everywhere. In response, governments have committed unimaginable sums of taxpayers' money to bail out banks and stimulate economic recovery. No one yet knows whether such measures will succeed. In the meantime jobs are disappearing, companies collapsing and public services evaporating.

On the positive side, the election of Barack Obama as US president has brought new vitality into international relations. The Obama administration and the Democratic majority in both houses of Congress have transformed the US role in climate policy, both domestic and international. Although US climate skepticism remains influential, intractable and deeply entrenched, the US response to climate concern is now creative rather than destructive, a change long overdue.

Political developments elsewhere are less encouraging. North Korea, now with nuclear weapons, is ominously unpredictable. Relations between Russia and western countries are edgy and uneasy. Domestic unrest has shaken China, Pakistan and other countries. More than two dozen countries have had food riots. The number of 'failed states' without effective government of any kind is increasing. Many millions of refugees and other poor in many places are living in squalor, without clean water, sanitation or electric light.

'Energy security' has become a key issue, despite deep disagreement about its meaning and implications. Confrontations between Russia and Ukraine over unpaid bills have cut off supplies of natural gas to parts of Europe in mid-winter, interrupting both heating and electricity. Concern over dependence on potentially unreliable gas imports has prompted a resurgence of conventional coal-fired electricity generation, despite its deleterious effect on climate. Some policy people acclaim and advocate trapping carbon dioxide emissions from coal and injecting them deep underground – so-called 'carbon capture and storage' or CCS. The technology appears feasible, if expensive; but it has yet to receive much more than rhetorical support anywhere.

Nuclear power, too, has climbed up the political agenda, with intensive talk of a renaissance after more than two decades of stasis. Thus far, however, talk has not been accompanied by much actual action. The first new nuclear plant ordered in Europe for many years, Olkiluoto 3 in Finland, has seen its cost increase from 3 billion to 4.5 billion euros. Within three years from start of construction it has fallen three years behind schedule. The follow-up plant, at Flamanville in France, has encountered similar problems. Doubts about capital costs, waste management, electricity prices and other risks continue to keep private investors acutely wary of the nuclear option. In Europe and North America advocates continue to press for government support. Elsewhere, however, the only nuclear plants under construction, such as those in China, are decreed by governments, with government funds.

Electricity from renewable sources, especially wind, is expanding rapidly in many parts of the world. Like other technological developments, however, it suffers from the general financial malaise. So do the extensions, upgrades and modifications of networks that are a prerequisite for major contributions from wind and other renewable electricity. The Obama administration and European groups are pressing for the establishment of 'smart grid' technology; but progress is slow. The potential of solar thermal generation south of the Mediterranean, with links into Europe, is exciting enthusiasm. But the scale, cost and risks of such projects remain daunting.

Interest in small-scale local generation, 'making your home a powerplant', is steadily increasing, boosted by so-called 'feed-in tariffs' that pay a premium price for the electricity produced. Perhaps the key corollary of local generation is the accompanying recognition that user-technology, including the built infrastructure, is a crucially important part of electricity systems. Buildings and other user-technology are often the part of the system for which performance is easiest to upgrade. The Obama administration in the US, Professor Nicholas Stern in the UK and many other commentators have argued that proposed economic stimuli should focus on upgrading electricity user-infrastructure and technology, particularly that used by governments themselves.

With all this in mind I reread *Keeping The Lights On*, expecting to find many passages needing updates. I was wrong. Significant recent developments do not invalidate the analysis in these pages, neither in general nor in detail. On the contrary: many of the most encouraging developments strongly reinforce the arguments presented here, bringing them more and more into the mainstream of policy and planning. Indeed, since completing *Keeping The Lights On*, I have been exploring the implications of extending the same approach beyond electricity, to embrace all human use of energy in all its forms. The project, called 'Managing Energy: for climate and security', is now well under way, under the joint aegis of Chatham House and the Sussex Energy Group. See Walt Patterson On Energy, www.waltpatterson.org/news.htm, for the latest information.

Many years ago the legendary US cartoonist and commentator Walt Kelly, through the voice of his protagonist Pogo, summed it up succinctly: 'We are confronted by insurmountable opportunities'. I hope we can respond in time.

Walt Patterson
Chesham Bois, Buckinghamshire
July 2009

ORIGINAL PREFACE

This is not the book I thought I was going to write. In the autumn of 1998, when I completed my book *Transforming Electricity* and delivered it to the publishers, the next stage of the work appeared to be straightforward. It would be a follow-up book called *Keeping The Lights On: Public Service in Liberalized Electricity*. In mid-1999 Chatham House published my briefing paper entitled *Can Public Service Survive The Market?: Issues For Liberalized Electricity*. It asked whether electricity could be at once liberalized and reliable, liberalized and universal, liberalized and sustainable. The answers, however, were far from obvious. Worse still, the questions too expanded with alarming speed. The longer I analysed them the more difficult they became. I soon realized that I did not understand either the questions or the answers well enough to produce a book I'd want my name on.

Meanwhile, as my colleagues at Chatham House waited patiently, electricity issues were evolving at a breakneck pace. Electricity decision-makers, swept along by events, seemed unlikely to take time to read an entire book. Instead I drafted a succession of shorter pieces, partly to disentangle the issues for myself and partly because shorter pieces, published immediately on the Chatham House website, might actually find readers. Every now and then I tried again to pull the analysis together into book form; but it was still too fuzzy.

Then, in 2005, to the astonishment and bemusement of many, nuclear power re-entered the policy agenda. Politicians, journalists and even environmentalists, knowing no history, listened to nuclear promoters and accepted recycled arguments that had been comprehensively demolished two decades earlier. The thought of starting the whole debate all over again numbed my brain. But my wife Cleone took a more practical view. She pointed out that I had devoted the 1970s and 1980s to an exhaustive refutation of the purported case for nuclear power – that the necessary evidence was right there on my study shelves. Within a few weeks she had scanned and digitized four books and a lengthening catalogue of other material. In February 2006 we launched a website archive of some 35 years of work – Walt Patterson On Energy, www.waltpatterson.org. The website archive included extensive commentary on nuclear issues, some dating back to the 1970s, still often dismayingly valid, indeed more so. By the end of 2006 it had received well over 100,000 hits, from more than 60 countries. But the site also included a sequence of analyses and presentations on energy and electricity in society, with what seemed to me profound implications for reliable, equitable and sustainable services. The successive pieces tracked my evolving understanding from first principles into some fascinating, radical and exhilarating territory.

Surveying the growing archive, I realized that I could now present the entire narrative between two covers, telling a coherent story I could call *Keeping The Lights On: Towards Sustainable Electricity*. That is what you are now holding. I hope it makes as much sense to you as it does to me. If it does, please help us to make sure our decision-makers hear about it. As we grapple worldwide with looming threats to climate and energy security, we don't have much time to get this right.

Walt Patterson
Chesham Bois, Buckinghamshire
January 2007

ACKNOWLEDGEMENTS

My colleagues at Chatham House in London have been very patient. During the gestation of *Keeping The Lights On* I went from being senior research fellow to retirement to associate fellow, in what was the Energy and Environment Programme, became the Sustainable Development Programme and is now the Energy, Environment and Development Programme, a span of more than seven years. I am grateful not only for their forbearance but for their support and encouragement. My thanks in particular go to Head of Publications Margaret May, to Programme Heads Duncan Brack and Richard Tarasofsky, to senior research fellow Valerie Marcel and to Programme staff Lorraine Howe, Gemma Green and Inge Woudstra-van Grondelle, who never once asked me, as they well could have, 'Are you ever actually going to write this book?'

Jonathan Sinclair Wilson and Tamsine Green of Earthscan gave me the opportunity to pull the story together and put it between two covers, my first-ever title not only in hardcovers but with a dust jacket; my warmest thanks to them both, and to Bob Faherty of the Brookings Institution, the US distributor.

Many colleagues around the world, some listed in Annex 4, 'Further Information', are now engaged in developing these exciting ideas. I'm grateful for the stimulation and insights they are providing. In particular, to Keith Barnham, Bill Frost and Becky Willis, who read drafts and offered valuable comments, my thanks.

Putting together our website Walt Patterson On Energy, www.waltpatterson.org, provided the impetus to prepare the book you're now holding. GreenNet, www.gn.apc.org, our long-time hosts for email and internet access, made the website almost effortless. For their unfailing support and cheerful reliability we thank them heartily, and recommend their services without reservation.

My long-time assistant Karen Lawther continues to keep me from disappearing under source material – my warmest thanks to her.

My family learned at last that when I wander about the house with a furrowed brow, ignoring them and bumping into things, I'm not angry or sulking – I'm writing. They put up with my lowering presence with remarkably good grace. Their endless forbearance, staunch support and buoyant companionship is my astonishing good fortune. To our delightful daughters Perdy and Tabby, and to my beloved Cleone, my gratitude and my love.

To Cleone, who lights up my life

Introduction

TOWARDS SUSTAINABLE ELECTRICITY

Electricity should be easy. Of all the urgent energy issues we face worldwide, electricity offers the richest possibilities. Yet we seem determined to choose the toughest, the slowest, the riskiest. Can't we do better at keeping the lights on?

The answer of course is yes. But we need to understand what we're doing. What we think we know about electricity is now obsolete and dangerous. What we need to know is still emerging, disconcerting but exciting. This book tries to show how our understanding might evolve. Starting from everyday experience, challenging traditional assumptions, it proposes an innovative approach that offers us abundant opportunities to upgrade our use of energy, especially electricity.

Part 1 sets the stage. It describes what we mean by 'energy', what we might mean by 'sustainable energy', and how we might go about 'Making Energy Sustainable'. 'Making the World Work' discusses what we want from energy, how we get it, and how we might do it better. 'The Energy Dilemma' revisits the reasons why this matters. 'Rethinking Energy' offers a starting point.

Part 2 sharpens the focus, from energy in general to the particular form of energy we call electricity. It shows how readily we might go about 'Making Electricity Sustainable'. 'Full Circle' suggests we might start by recalling Thomas Edison's original business plan. 'The Electric Challenge', 'Generating Change' and 'Networking Change' explore the implications for electricity systems, generation and networks. 'Decentralizing Networks' pursues a key theme in more detail.

'Getting the Story Right' proposes a new narrative for electricity, more accurate and more useful than the traditional story we've been telling for a century. 'Getting Energy Right' reiterates key ideas for emphasis and takes them further, showing how getting electricity right might help us to tackle broader energy issues, including energy security and climate change. 'Sustainable Electricity: Changing Minds' describes how we can start making electricity sustainable by changing the way we think about it.

Annex 1, 'Running the Planet', surveys the larger context within which we live. Annex 2, 'Discussing Energy: A Style Guide', is a guide to how we talk about it, how we get it wrong and how we could get it right. Annex 3 is a glossary of terms used in this book, from electric jargon to the many generating technologies now available. Annex 4 points to further information.

Transforming electricity will not be easy. But it will be essential, and profitable. It will be a key to energy security and climate security. If we do it right it might even be fun.

Part I

MAKING ENERGY SUSTAINABLE

1

MAKING THE WORLD WORK

This is embarrassing. Many years ago I wrote a book called *Energy and Purpose*. 'Purpose' meant 'what we humans want from energy, and how we try to get it'. A reputable publisher gave me a modest advance. I worked on the book for two years, eventually accumulating about 100,000 words of text. But the longer I worked on it the less I liked it. I finally had to confess to myself that I didn't know what I was talking about. I didn't understand enough about energy and purpose to say anything really useful or persuasive. I gave the publisher back the advance, piled the typescript in a cardboard box and stashed it in the archive, along with the unpublished novel, the unpublished textbook and the unproduced musical.

In the coming pages I am returning to the scene of my failure so long ago, to write again about energy and purpose – what we humans want from energy, whether we can get it and, if so, how. I have an alarming sense of déjà vu, knowing I have been here before, and wondering whether I can do any better this time. Wish me luck.

Why write about energy and purpose? The short answer is that we're making a mess of it. The world isn't working well enough. More than two billion people, one-third of humanity, have no access to the kinds of energy benefits the rest of us take for granted; and the proportion of 'energy have-nots' is increasing, not decreasing. Worse still, the key fuels and energy technologies of the 'energy haves', like us – fossil fuels, large dams, nuclear power – all face problems that look insuperable. Worst of all, doing what we do with energy is disrupting the climate of our only planet. If that doesn't worry you, it should.

What do *you* want from energy? You probably never thought about it. That's as it should be. Almost everything you get from energy you

This chapter is adapted from Institute of Energy Melchett Medal Lecture, London Planetari
22 June 2000.

get without even noticing. It doesn't involve a meter; you don't get billed for it. You get surroundings whose temperature mostly stays within limits your body can tolerate. You get sunlight processed by green plants, which store up the solar energy in a form you can eventually eat, for your muscles to use. As a by-product from the green plants you get the oxygen you breathe to process the food; and so on. You are immersed in, indeed you are a part of, natural energy systems of astonishing complexity and variety; and you take them all for granted.

You are also, however, immersed in energy processes that you yourself, and other people, initiate and control – what we can call human energy systems. Some you take for granted as completely as you take natural energy systems for granted. You probably can't remember the last time you turned on a light. Some human energy systems you notice, at least some of the time, particularly when they fail. When you turn the key in the ignition, or flip the light switch, and nothing happens, you notice. You also notice when you get a bill. That may be part of the problem. In the past three decades we have come to think of energy as something you get a bill for. That must change.

Using energy

Start with this word 'energy'. When you think of energy, you probably think of oil, coal, natural gas, electricity. You shouldn't. The language we now use to talk about energy is not just wrong – it's actively misleading. If we can't even describe the issues and options correctly we'll never get the policy right. How many times have you heard or read some energy specialist refer to 'energy production' or 'energy consumption'? These people are supposed to be experts. Surely they ought to know one unbreakable law, the First Law of Thermodynamics, the law of conservation of energy. *No one* produces energy. *No one* consumes energy. The amount of energy in *the whole universe* remains the same. That's what makes energy such a valuable and important concept to understand how the world works. We don't have to conserve energy. Nat[illegible] it for us.

[illegible], do we talk this way? The answer is simple. When we talk [illegible] production, energy consumption and energy conserva[illegible] mean 'energy'. We mean 'energy carriers' – that is, fuels [illegible] The confusion dates back less than 40 years. Until the early 1970s governments had 'fuel policy'. They had Ministries of Fuel, or perhaps of Fuel and Power – 'power' meaning electricity. Then, in October 1973, the Organization of Petroleum Exporting Countries (OPEC) suddenly quadrupled the world price of oil, plunging the world into a panic. Governments everywhere launched a frenzied search for a 'substitute' for oil. Within weeks all the different fuels, plus electricity, were swept together and called 'energy', as if they were all potential substitutes for one another, all more or less interchangeable. 'Fuel policy' became 'energy policy'. Governments exhorted their citizens to 'conserve energy'. Ministries of Fuel became Departments of Energy. Oil companies, coal companies, gas companies and electricity companies all became 'energy companies'. In the UK the Institute of Fuel became the Institute of Energy; much the same happened all over the world.

But of course everyone knows that specialists talking about 'energy' really mean 'energy carriers' – oil, coal, natural gas, electricity. Lumping them all together and calling them 'energy' is just a convenient shorthand. Does this quirk of language really matter? Yes, it does. It distorts our understanding of what we are actually doing with energy; that is, 'energy', not 'fuels and electricity'. Worse still, this misleading language obscures crucial options we now have, ways for us to use energy much better.

Note that phrase, 'using' energy. That's what we do with energy. We don't consume it, we use it. Humans have been using energy on purpose since long before the beginning of recorded history. Our human ancestors began using energy by intervening intentionally in natural energy flows, or what we can call 'ambient energy' – energy that is there for us to use, with no meter and no bills to pay. The first 'energy technologies' that our human ancestors hit upon were clothing and shelter. In cold weather, clothing reduces the loss of heat energy from your body; in hot weather, it protects you from too much solar energy. Shelter provides an enclosed space, reducing energy flows and keeping the temperature inside more stable than that outside; inside the shelter you are more comfortable. You may not usually think of clothing and shelter as energy technologies. But if you really want to understand how we humans use energy, clothing and shelter are fundamental. Note, too, that clothing and shelter are physical materials. You don't measure or pay for the energy flows involved. The clothing and the shelter manage the energy flows for you.

Humans were probably manipulating ambient energy in these basic but fundamental ways long before they learned to control fire and use fuel. Fire and the fuel to feed it opened many new possibilities. Nevertheless, intervening in ambient energy remained an important aspect of using energy on purpose. In many parts of the world, for instance, humans developed increasingly subtle and ingenious ways to design the energy technologies we call buildings. They selected materials and erected structures to use the ambient energy of sunlight, of moving air and human bodies to deliver comfort, light and ventilation. They also developed technologies including sails, windmills and watermills, to use the ambient energy of wind and water for human purposes.

Ambient energy is all around us, whether or not we explicitly want to use it. Fuel, by contrast, is a material containing energy that we can release on purpose, when and where we want to use it. The word 'fuel' comes from old French 'fowaille', which comes in turn from low Latin 'focale' and Latin 'focus', meaning 'fireplace'. Etymologically, a fuel is 'material for a fireplace'. Historically, a fuel is a material you can burn, to release its stored energy as heat. This creates a local high temperature, in which you can cook food, fire ceramics and smelt metals. But the real potential of fuel only emerged less than three centuries ago, with the invention of the steam engine.

The steam engine could convert the heat energy from a burning fuel into mechanical energy – a source of controlled force and motion much more powerful than human or animal muscles, and more predictable than wind or water. The steam engine tipped the balance. Since the advent of the steam engine, giving us this potent additional way to use fuel energy, we have gradually forgotten about using ambient energy. Instead we have concentrated our attention on fuel energy – usable energy stored in a form that can be stockpiled, transported, and released in concentrated form, when and where we want to use it.

Note one important corollary. Fuel energy is comparatively easy to measure and quantify – so many tons of firewood or coal, barrels of oil, cubic metres of natural gas. Because it can be stored, it can be possessed – someone can take title to it and own it. It can therefore be bought and sold. Nobody can buy or sell ambient energy, because nobody owns it – not yet, at any rate. Keep the distinction between ambient energy and fuel energy in mind. It's important.

Energy technologies

The steam engine, and all the numberless energy technologies that have come after it, also demonstrate another key point. At its simplest, fuel energy can be released directly from the fuel and used as it comes – say from a bonfire. However, precisely because it is being released intentionally, for a human purpose, fuel energy is usually released in the context of some sort of physical hardware, an energy technology designed to control and direct the conversion of the fuel energy.

For example, my wife and I have a little house on a remote hillside on a Greek island we've been visiting since the 1960s. The house is heavily insulated, roof, walls, windows and floor, in order to take maximum advantage of the ambient energy, whatever the temperature outside, to keep us cool in summer and warm in winter. In northern Greece, however, winters can be cold. Rather than lighting a bonfire on the kitchen floor we have a black potbellied stove. It is essentially a metal canister with a lid, a small front door into which we put the fuel, and a pipe to channel the smoke of the fire out the back of the house. We burn dead heather branches from the hillside, scrap planks from the builders, cardboard packaging, essentially anything combustible. It converts the energy from the fuel into radiant heat energy that saturates the structural material of the house. If it's cold outside, a short burst of heat from the stove fine-tunes the temperature inside, and keeps us cosy for hours. Of course a lot of the heat from the stove escapes out of the chimney, and the emissions would probably get us into trouble in London. As an energy technology our potbellied stove could scarcely be more basic. But we have fallen in love with it.

Our potbellied stove, however, illustrates another significant aspect of human energy use. Precisely because the stove is such basic energy technology, it can use the most basic fuel – whatever we can lay our hands on to burn. The only processing the fuel requires is to break or cut it into pieces small enough to put in the stove. Although we bought and paid for the energy technologies we use, the house itself, and the stove in the kitchen, we don't have to buy the fuel. We can gather and cut it up ourselves. It costs us our own time and effort, but doesn't take any expertise.

In that respect, our stove is no longer a typical energy technology, at least in the richer parts of the world. Over the past three centuries, the interaction between fuels and energy technologies has become ever more specialized. A particular technology requires a particular fuel, and vice versa. The specifications of both the technology and the fuel have become steadily more stringent. Your car engine probably demands not petroleum, not even plain 'petrol', but unleaded premium petrol. A cooker designed for natural gas will not work safely on bottled propane; and so on.

That's the main reason why looking for a 'substitute for oil' in the 1970s was misconceived. You can't change the fuel without changing the energy technology that uses it. Preparing, delivering and supplying fuels appropriate for their corresponding energy technologies now requires not only high levels of expertise, but elaborate organization of all the necessary skills and competences, with all that that implies. You can't collect the fuel on a hillside. Just as you buy and pay for the energy technology, you also have to buy and pay for the fuel. The companies you buy the fuel from used to be similarly specialized – oil companies, coal companies, gas companies. That, however, is now changing rapidly. Companies that used to be oil companies or gas companies are now both. Companies that used to supply gas or electricity now supply both. Companies that were local or regional are now national and international. Companies that used to deliver gas through pipes and electricity through wires now own and operate both networks, often together, often across national borders. They now call themselves 'energy companies'. But their business is still mainly just fuels and electricity, which they must nevertheless manage distinctly and differently, however much they interact. If and when they become true energy companies, that will be another story, to which we shall return.

Energy systems

Within the past century, the human use of energy in much of the world has come to depend not merely on separate individual fuels and technologies, but on entire intricate human energy systems, complex and interconnected. To fulfil our many purposes these human energy systems use a combination of ambient energy and fuel energy not merely in individual energy technologies but in a vast human energy infrastructure. Enormous aggregations of buildings are expanding into megacities. The buildings are filled with other energy technologies, and linked by roads, pipes, cables and other interacting connections, extending human energy processes not only across entire continents but even bridging the oceans.

As well as natural energy systems, we now have a human energy infrastructure that also covers the planet. Much of this human energy infrastructure – buildings, appliances and other equipment – delivers the energy services we all want, such as comfort, cooked food, illumination, motive power, refrigeration, information and communication. However, a substantial part of this infrastructure – oilfields, pipelines, power stations and so on – is now devoted to collecting, preparing and delivering fuel energy to run the rest of the infrastructure. Making substantial changes to the delivery infrastructure can take as long as making substantial changes to the energy-service infrastructure, and cost at least as much.

Among the specialized and complex energy systems we have created, perhaps the most specialized are those that function with an energy carrier quite different from fuel: electricity. No matter what you hear from politicians and others, electricity is not a fuel. A fuel is a physical substance. You can store it until you want to use it or sell it. Electricity is not a physical substance. It is a physical process, happening simultaneously throughout an entire interconnected system. It has to be generated more or less exactly as it is being used. Fuels and electricity also differ in another fundamental way. A fuel such as natural gas comes out of a hole in the ground at a particular place. If you want to use it somewhere else you have to carry it there. Electricity, by contrast, you can generate anywhere. Just ask the person with the hissing headphones sitting next to you on the bus.

But electricity by itself is useless. Electricity just carries energy; the energy has to be converted into a useful form by an energy technology such as a lamp, a motor or a computer. When it is being used, the energy technology involved – lamp, motor, computer – becomes a functioning part of the electricity system. The whole system is part of the human energy infrastructure, operating continuously in real time. You can keep a stack of wood, a pile of coal, a tank of oil or even a canister of compressed natural gas on site, ready to use when you wish. But if you want to use electricity, the whole system has to be operating with you, in real time.

In that respect, oddly enough, using electricity has a lot in common with using ambient energy, and the link is going to get steadily closer. Like electricity, ambient energy is delivered continuously. To use ambient energy on purpose, you need physical assets – a building, a water turbine, a wind turbine, a photovoltaic panel – that is, physical infrastructure. For some purposes – such as comfort, probably the single most important human purpose for using energy – if you make the physical infrastructure good enough, ambient energy may well suffice, with no resort to fuel energy. In much of the world, however, we have accumulated a built infrastructure whose performance with ambient energy all too often seems wilfully poor, making fuel energy essential if we are to get the comfort we want.

When I first arrived in the UK from Winnipeg in Canada more than four decades ago, I could not believe the buildings in the UK. The heat inside barely slowed down before it escaped outdoors. We also settle for poor performance from the energy technologies inside and around the buildings. When first in London I lived in a bedsitter in Bayswater. The bath was in a sort of greenhouse over the front door. The boiler was in the basement. The hot water pipe from the boiler ran up the exterior wall. Not only was it not lagged, it was painted black, the best colour for radiators. The water running into the tub was barely tepid. The arrangement was so futile they need not have bothered.

That may sound like an extreme example, but it's not. Countless reports and analyses have underlined the inadequate performance of lighting, motive power and other energy technologies in many parts of the world, and deplored the missed opportunities for so-called 'energy efficiency'. Many reasons have been suggested. Meticulous and detailed lists of 'barriers to energy efficiency' have been pouring out since the 1970s, and they are all true. But the single underlying reason why our human energy infrastructure does not perform better is that most of us can't be bothered. We have other things to think about. If we are ever going to make the sweeping improvements in human energy use long since readily available, if we are ever going to make the world work better, someone has to want to do it. We have to find ways to make those who can do it want to do it.

Energy and money

What about costs? Energy itself costs nothing. However, if you want to use ambient energy you have to design and fabricate the technology to do so. If you want to use fuel energy you have to produce and process the fuel, and deliver it to where it is to be used; and you have to design and fabricate the technology to use the fuel. Once you get beyond the mud hut and the bonfire, all these activities become variously part of a financial economy, carried out in transactions mediated by money. The skills, competences, responsibilities and risks involved have been divided up and apportioned out in ways that once appeared to make sense, but now look profoundly unsatisfactory, because of what they have done to human energy infrastructure.

Consider, for instance, the two parts of this infrastructure. One part delivers the energy services we want. The other part delivers the fuels and electricity to run the first part. Both parts represent investments in physical assets. Because the fuels and electricity are to be sold by the unit to users, governments generally treat investment in all of this part of the infrastructure favourably for tax purposes, as business investment. If a company invests, say, in a new power station to sell electricity, the government allows it to claim tax relief. If, however, we all invest, say, in more efficient refrigerators, to make the new power station unnecessary, we get no tax relief on our investments. This one single anomaly, replicated across all the energy infrastructure, skews the pattern severely in favour of more investment to deliver fuels and electricity, and less investment to deliver better energy services. Traditional tax regimes encourage investment in infrastructure that makes money, rather than in infrastructure that delivers the energy services we citizens want.

Using ambient energy does not make money – not at the moment. But fuel energy can be stored and sold, by the unit. What costs money is not the energy, but storing it, carrying it to where it is to be used and converting it. We use fuels and electricity to have energy available where, when and in what form we want; and we pay for the privilege. Policy people call this 'commercial energy', as if paying for it in some form of 'market' makes it better. Commentators scrutinize the prices of fuels and electricity, and analyse their movements minutely. However, in our modern interconnected society the prices of fuels and electricity by the unit have long been essentially artificial, shaped by preferential tax regimes, subsidies and cross-subsidies, cartels and outright monopolies, as in the case of electricity networks. With this in mind the highly respected chairman of Ireland's Electricity Supply Board, Patrick Moriarty, once remarked succinctly, 'The price of electricity is what the government wants it to be.' What with taxation, subsidies and other interventions, much the same can be said of fuels. Except for short-term advantage, unit price is not a good enough criterion.

'Sustainable energy'?

If we were stuck with these traditional arrangements for using energy on purpose, concentrating on selling fuels and electricity by the unit at more or less arbitrary prices, we would have little chance of making the world work better. But we have another option, as following chapters will describe, an innovative approach to counter the pessimism that says we must use more and more fuels no matter what that does to the planet.

For years we've been talking about 'sustainable energy'. Following the lead of the landmark Brundtland Commission report *Our Common Future*, published in 1987, we could define 'sustainable energy' as 'energy use and supply that meets our needs without jeopardizing the ability of our children to meet theirs'. Unfortunately, we can see all too clearly that the way we now use and supply energy is jeopardizing the future, not only for our children but for the entire ecosystem of our planet. Even on its own terms it looks acutely vulnerable, with long and tenuous supply lines precariously easy to disrupt, for reasons of money, politics or malevolence.

As we scramble for 'energy security' and strive, all too belatedly, to get a grip on climate change, we may at last initiate the energy evolution that will bring us sustainable energy – not just those of us among the 'energy haves', but also the two billion still waiting. If, against the odds, we somehow get this right, in the course of this century we may even manage to make human energy systems work more like natural energy systems, continually delivering the services we want while most of us don't even notice. We now understand a great deal about human energy systems; but natural energy systems still remain in key respects tantalizingly beyond us. The American energy visionary Amory Lovins once offered a vivid illustration. We know, he said, three ways to make a building material out of limestone. We can cut it into blocks. We can calcine it in a furnace to make cement. Or we can feed it to a chicken. Weight for weight, eggshell is one of the strongest materials we know. But we don't know how the chicken does it. What's more, the chicken does it not in a furnace but at its own body temperature. As yet, we humans manipulate materials and derive other energy services mostly using substantial temperature differences, in processes that use brute force rather than elegance, especially by burning fuel. Can human energy systems converge towards natural energy systems? The vision is appealing and exhilarating.

Energy, to be sure, is only one of the fundamental issues that challenge us. But if we don't get energy right the other issues will be insoluble. To herald the new century, Chatham House published my millennium essay 'Running the Planet', an attempt to reassess the fundamentals of human life on Earth from first principles; you can find it in Annex 1. As the title indicates, whether we like it or not, we humans are now in charge. Our future is up to us. But we cannot long survive as a species, on this interconnected planet we share, unless we can rectify the gaping disparities that divide us: 'The co-existence of opulent luxury and desperate poverty, sometimes within the same urban area, is not a recipe for stability.' Nor can we keep borrowing from our descendants.

If we are to meet this challenge, we have to get energy right. We have to make the world work better, and we can. But it will not be easy. In the closing lines of 'Running the Planet':

> *No one knows all the answers. We may not even be asking the right questions. We are all in this together, and we'll need all the help we can get.*

2

THE ENERGY DILEMMA

Strike a match.

You have just done something uniquely human. No other living creature, so far as we know, can start a fire at will. When our ancestors learned to control fire, they took the first step towards becoming the dominant form of life on Earth. Controlling fire was the key to using what we now call energy. If you can control fire, you can keep your body comfortably warm wherever you are, whatever the time of day, whatever the season. Instead of eating food raw you can cook it, vastly extending the edible menu. You can bake watertight pots and durable bricks. You can extract iron and other metals from the earth, and work them into strong and functional tools.

We have come a long way since the first intentional fire threw dancing shadows on the walls of a cave. We now know how to burn not only wood, dung and fat from animals and plants, but also coal, petroleum and natural gas. We can capture usable energy from moving water and blowing wind, and even release it from uranium. We can convert energy through incredibly intricate processes, at every scale from gigantic dam to silicon microchip. By using energy, we humans have transformed much of the natural world to suit ourselves.

We do not, however, know as much about using energy as we thought we did. With all our experience we still keep getting it wrong. An old adage is catching up with us: playing with fire can be dangerous. Indeed it is now endangering our entire planet. We need to re-examine urgently this unique human ability, the ability to start a fire – that is, the ability to use energy on purpose. If we do not, it may soon be terminally out of control.

Think about your burning match. What could you do with it? You could light a bonfire to bake your potatoes; or you could raze a forest.

This chapter is adapted from first publication in *The Energy Alternative: Changing the Way the World Works* (Boxtree, 1990, Optima, 1991).

You could ignite your kitchen gas-ring; or you could level a shopping centre. You could fire up the coal in a factory boiler; or you could set off a disastrous explosion in a mine. Hazards like these are essentially immediate and local. People have lived with them for centuries. We try to minimize them; but we accept them in exchange for the obvious benefits we gain by using energy. Within our lifetimes, however, we have become acutely aware of other hazards of energy use, more subtle and indirect, more long-term and far-reaching. Choking smog from car exhausts and coal fires turns urban air a dirty orange; eyes water and lungs inflame. Acid rain laden with sulphur and nitrogen oxides from burning coal and oil poisons rivers, lakes and soil; fish die and woodlands wither. The explosion of the Chernobyl nuclear plant showed that even a single accident can affect geographical areas thousands of kilometres away, for decades to come.

Now, however, all these unintended consequences of energy use, alarming though they be, have been overtaken by yet another – the most awesome yet. Scientific evidence leaves no room for doubt. We are now upsetting the climate of our planet. Since the turn of the new century we have had a succession of the hottest years ever recorded, bringing droughts, violent storms, floods and extremes of weather to many parts of the Earth. Glaciers melt; rivers vanish. Leading scientists concur that we are now seeing the so-called 'greenhouse effect' advancing with alarming speed. Gases we have released into the atmosphere are behaving like a greenhouse surrounding the planet. When the sun shines into this greenhouse it raises the temperature inside, producing unpredictable changes in complex planetary systems such as the atmosphere and oceans. Using energy the way we do is a key cause. One of the main 'greenhouse gases' is carbon dioxide, released by burning coal, oil or natural gas.

Every time you strike a match, turn on a light or press the starter on your car you complicate matters that little bit further. Your impact alone is negligible – but you are one of more than six billion people on Earth. Countless millions of your fellows are also turning on lights, pressing starters, and using energy in an endless variety of other activities, in households, vehicles, offices, factories and wherever else they happen to be. To be sure, at least two billion people have little opportunity to use energy the way we do in industrial countries; but they can claim with justice that they are entitled to share the consequent benefits the rest of us now enjoy. If all six-plus billion of us, and the additional billions yet to come, try to use energy the way we now do in industrial countries, the outcome will be catastrophe. We urgently need to rethink what we are doing with energy.

Thinking about energy

Until recently, most of us fortunate ones in rich countries have scarcely thought about energy at all. As long as the lights stay on and the car starts, energy is someone else's responsibility, someone else's problem. Even when we do think about it, we don't think about 'energy'. We buy petrol or diesel for the car. We pay the gas bill and the electricity bill. We users actually know better than to lump all the different fuels together with electricity and call them 'energy'. We know instinctively, intuitively, that the different fuels are not interchangeable, and that electricity is yet more different. But we don't think about it. We just take for granted that when we buy a car or a computer, a lamp or a boiler, someone will ensure that we can also get the fuel or electricity we need to run it.

In practice we expect governments and companies to do the necessary thinking for us. Many millions of people around the world make energy their business. Some design, manufacture or sell the technologies that help us use energy. For technologies that use fuels or electricity, some produce and deliver these energy carriers. Some analyse resources, prices and patterns of energy use and supply, to anticipate and plan for future developments. Whatever their particular involvement in the energy business, they have had to devise manageable ways to think about it. Never before have they had to think as hard as they do now. The conclusions they reach and the decisions they take affect every aspect of your life, from the bills you pay to the air you breathe.

Governments establish ground rules – taxes, environmental regulations, planning laws – that attempt to reconcile the need to protect climate, air, water and land with the need to keep the lights on. Companies try to anticipate the energy hardware we shall want to use, the fuel and electricity supplies they will require, and how much they will cost, to channel investments in the right direction. International organizations such as the United Nations, the World Trade Organization and many others, global and regional, try to mediate between different nations whose interests often conflict. If one national government imposes more stringent environmental controls than another, it may find that its manufactured goods become more expensive than those of less environmentally scrupulous competitors. On the other hand, if international environmental standards are lax enough to suit the least scrupulous country, the climate bears the brunt. The quality of air, water and land suffers. So do those who breathe, drink and live on them. Pollution and greenhouse gases pay no attention to political borders. Why should one nation clean up its act if its neighbours do not? The atmosphere and the oceans are 'commons', accessible to everyone. If we treat them as communal rubbish dumps we all suffer the consequences.

The energy and environmental policies that governments, companies and international organizations adopt and implement have a direct and immediate effect on their citizens and their customers. Government, corporate and international policies influence the kind of homes, cars and consumer goods you can purchase, and how much they cost to buy and to operate. They determine how much you pay for fuels and electricity. Above all, the policies affect the health and vitality of the climate and environment in which you live, both local and global.

For half a century, at least in rich countries, the traditional way of thinking about energy has served most of us moderately well. Until the 1970s, governments, fuel companies and power companies made what they called 'forecasts' of future requirements, and planned their investment programmes to supply the requisite fuels and electricity accordingly. The forecasts proved to be more and more inaccurate, so much so that governments and companies stopped calling them 'forecasts'. They still planned ahead, but did so on the basis of 'scenarios' or 'projections'. But the aim was, and is, the same – to guide investment in facilities to produce and deliver coal, oil, natural gas and electricity, which they continue to call energy. The process was difficult, and grew steadily more so. Since the turn of the century it has become effectively impossible. Reconciling the colliding contradictions is now beyond the power of what governments and companies have traditionally called energy policy.

Why is this so? Start with expectations. In countries that have been rich for at least half a century we still expect to use increasing amounts of fuels and electricity year by year, essentially indefinitely, indeed we are doing so. Meanwhile, around the world, other countries, among them China and India, with a third of the people in the world, are rushing to catch up, following the traditional energy patterns of the past two centuries, focused on fossil fuels. The scramble for access to fuel resources, especially oil and gas, is already intense, and intensely politicized. The Organization of Petroleum Exporting Countries tries to close valves and cut production, to keep the oil price high. Russia's Gazprom flexes its massive muscles internationally, shutting off gas supplies unilaterally and demanding doubled or quadrupled prices, and overturning agreed contracts to demand a majority share of joint international projects. China is now a major player in oil and gas provinces around the world. Pipelines from oil- and gasfields in central Asia may soon flow not west but east. As new buyers flock into the global market for hydrocarbon fuels, the rich countries of Europe and North America face a daunting challenge to what they call 'energy security'.

A growing number of experts even believe that we may be approaching what they call 'peak oil', the peak of petroleum production worldwide, after which the world will no longer be able to replace the oil reserves we have produced and used. If this belief proves to be well founded, it does not mean that oil will run out; but it does mean that oil will become progressively more expensive. Since the nineteenth century the twin attractions of petroleum as a fuel have been its versatility and its surprisingly low cost. These attributes have given modern industrial economies an addiction to oil. If they have to kick the habit they can expect to suffer severe withdrawal symptoms.

Other hydrocarbon options are available. They include so-called 'oil shales' and 'tar sands' that can be processed to produce expensive equivalent barrels of fuel. They may also include so-called 'clathrates', deeply buried and exotic potential sources of methane. Above all they include the commonest fossil fuel, coal, which can be 'hydrogenated' to produce liquid fuels. Unlike petroleum and natural gas, coal remains abundant, widespread and cheap. China, India and many other countries, including once again the US, expect and intend to use more and more coal in the years to come, especially to generate electricity.

That, however, now looks like the final ingredient in a recipe for planetary disaster. Pouring yet more fossil carbon dioxide into an atmosphere we have already destabilized is foolhardy going on suicidal. The world's leading climate scientists, the Intergovernmental Panel on Climate Change (IPCC), concur that the Earth's climate is already changing, and that emissions of carbon dioxide from fossil fuels are a significant cause. Initial analyses, beginning in the 1980s, were enough to persuade the world's governments to sign and ratify the Framework Convention on Climate Change (FCCC), at the UN conference in Rio in 1992. By 1997 the evidence was becoming sufficiently alarming that most governments signed and ratified the so-called Kyoto Protocol to the FCCC, undertaking legally binding commitments to reduce their emissions of carbon dioxide and other greenhouse gases. Unfortunately the world's largest emitter, the US, rejected the Kyoto Protocol. US politicians and corporations denied that climate change was a serious issue; many called it a hoax. The US administration under President George W. Bush declared that imposing limits on the use of fossil fuels would damage the US economy. Now, however, the cumulative evidence is mounting at appalling speed, as even many across the US acknowledge. The deniers are still there, but their protests are fainter. In late 2006 the former chief economist of the World Bank, Sir Nicholas Stern, presented a study concluding that the potential economic damage from climate change far outweighed any negative economic impact of palliative measures.

Putting the question

The question then becomes blunt and direct. We know that our greenhouse gas emissions are changing the climate. What can we do about it? There lies the dilemma, simple and brutal. Official and quasi-official projections and scenarios all declare not only that we shall continue worldwide to use fossil fuels, but that the amount we use will continue, essentially indefinitely, to increase, not decrease. That means in turn that carbon dioxide emissions will continue to increase, as will its concentration in the atmosphere. The Earth's average temperature will continue to rise. So will sea level. Polar ice will shrink, islands disappear. Permafrost will thaw, releasing more methane, a pernicious positive feedback accelerating the thaw. Droughts, floods and hurricanes will strike more often, more widely and more severely. Agriculture, habitat and wildlife will suffer. So will our children. The prospect is grim, and growing grimmer. We cannot reconcile climate security with what we have come to call energy security. The two are incompatible.

To be sure, many people are trying very hard to reconcile them, advocating an assortment of policy measures to mitigate the impact of carbon dioxide emissions. Everyone everywhere claims to be in favour of 'energy efficiency', whatever they may think that to mean. They have been so, at least rhetorically, since the early 1970s, with at best marginal impact on the way we use energy. A rapidly growing constituency around the world advocates much more development of various forms of so-called 'renewable energy', including wind power, solar thermal power, solar 'photovoltaics', biomass fuels and electricity, and various forms of marine energy from tides and waves. We shall have more to say about some of these in Chapter 4. A diehard band of promoters has seized upon climate change as the last chance to resuscitate the fading fortunes of civil nuclear power. They have won some eminent and surprising converts, to what still looks a profoundly risky proposition to anyone aware of nuclear history.

Some advocate changing the way we live, and some of the values many of us live by, including for instance the opportunity to travel long distances by air. They propose that an equitable distribution of the six billion tonnes of carbon dioxide that six billion people emit annually works out to one tonne per person per year. They invite governments to initiate a process of 'contraction and convergence', by allocating carbon allowances, whereby those who emit too much carbon buy allowances from those who do not. Government and corporate versions of such schemes are already under way in Europe and North America. Called 'emissions trading systems', they depend crucially on the initial allocations of allowances to be traded. Thus far the allocations have been too generous to impose much restraint on total emissions. The overall effect has been to deliver not climate protection but windfall profits to those receiving allowances. The allocation problem as it affects companies will be multiplied manyfold by any move towards personal allowances to individuals. The political feasibility of such measures in a democratic context still looks dubious at best.

Back to fundamentals

Must we therefore concede defeat, with no way out of this energy dilemma? Or are we all missing something here? Surely we do not have to choose between energy use and environment, between energy security and climate security. We are seeing a profoundly misleading picture of what is really happening. Before we concede that we cannot make energy use sustainable, we ought first to go back to fundamentals, to rethink exactly what we are doing. We need to be sure what we actually mean and intend when we use energy on purpose. Then at least we may be able to see what looks sustainable and what does not.

If you are an energy planner or policymaker, one point is clear: it is impossible to do everything. You can contemplate today an extraordinary shopping list of options – different technologies to use and supply energy; different financial arrangements, laws and regulations; different organizations and institutions. But you must choose among these options, and hope you are making the right choice. The same challenge arises at every level, from domestic to international. Do you select this deep-freeze, even though it costs more? Do you select this pollution-control standard, even though it costs more?

The choices are further complicated because no option is guaranteed trouble-free. Two decades ago you could have selected your new deep-freeze, happy in the knowledge that its operating system contained only the safe, inert chemicals called 'chlorofluorocarbons' or CFCs. Then we discovered that CFCs damage the protective ozone layer that shields the Earth from the sun's ultraviolet radiation. Even the most apparently innocuous choice we make may subsequently backfire on us, scarcely a reassuring realization amid so much uncertainty.

Yet never before have we been offered such a sumptuous shopping list of energy options. We can choose among different energy applications, their technologies and their suppliers; different fuels and their suppliers; different supply technologies and their suppliers. All the options have advantages and disadvantages, some evident and some much less so. We have both too much information and not enough. Advertising, official reports from governments and international organizations, unofficial reports from universities and environmental organizations, analyses from investment brokers, articles in print, broadcasts on TV and radio, blogs on the internet – keeping up with the outpourings is more than a full-time job; but major gaps remain unfilled.

As yet we simply do not know enough about the comparative costs and performance of different energy technologies, or about their environmental impacts, or about the feasibility of competing options. Enthusiastic salesmanship, whether for nuclear power stations or for wind turbines, is no substitute for actual experience, as we have all too often realized too late. Despite the uncertainties, however, we must make choices. Even to sustain our present patterns of energy use, someone would have to draft and implement policies and plans, to design, order, install and operate the requisite technology. Some commentators still expect us to do just that, for decades to come. As the climate threat looms, however, rational analysis leaves no doubt that our present patterns of energy use are not the solution but the problem, and long overdue for change.

If we are to reach the best available decisions, if we are to move towards sustainable energy, we must understand clearly what we know and don't know, starting from first principles. We must describe the opportunities and risks as accurately as we can, and act accordingly. As we pick our way among the energy options and uncertainties – scientific and technical, economic and financial, social and political – one key question becomes paramount. Who is to make the energy choices that will determine the sort of world we pass on to our grandchildren? Who decides, and how?

Think about it the next time you strike a match.

3

RETHINKING ENERGY

'Energy is eternal delight.' So wrote the English poet and mystic William Blake in 1790. After more than two centuries, however, the ecstatic purity of Blake's dictum has been more than somewhat muddied. To Blake, 'energy' was a manifestation of the vitality of the world – an essentially mystical concept. He would have recoiled from any thought of 'defining' energy, much less measuring it and describing it with numbers. But the 'dark satanic mills', the coal-burning factories that Blake deplored, were going to change 'energy' from 'eternal delight' to something much more problematical.

According to the Oxford English Dictionary, 'energy', the word itself, made its first recorded appearance in 1599, meaning 'force or vigour of expression' – a sense undoubtedly congenial to Blake. It is equivalent to the Greek word 'energeia', used by Aristotle, and comes from the Greek work 'ergon' meaning 'work'.

Other meanings gradually accrued. By 1665 it meant 'power actively and efficiently exerted', and by 1667 'ability or capacity to produce an effect'. The word 'energy' described an attribute recognizable by its consequences. You could tell that a person had 'energy' by watching the person in action. A person with energy could move swiftly, or lift heavy weights, and might get hot and sweaty doing so. You could even say that one person had 'more energy' than another; but the comparison was vague and imprecise.

We still use the word 'energy' like this in everyday speech; it conveys a meaning that is adequately clear and intelligible for everyday use. But the association of a person's 'energy' with movement, force, work and heat, is a clue to the much more potent and precise application of the concept of energy, which emerged just over 150 years ago.

This chapter is adapted from first publication in *The Energy Alternative: Changing the Way the World Works* (Boxtree, 1990, Optima, 1991).

Discovering energy

By the seventeenth century, even as the word 'energy' was coming into parlance, thinkers using a new approach called 'natural philosophy' – what we now know as 'science' – were seeking new and more powerful ways to understand and describe the world. One key idea was comparison not only with words but with numbers. Instead of simply calling one horse 'short' and another 'tall', you counted the number of your hand's breadths from the ground to its shoulder: one horse might be 14 hands high, another 15, another 16. Provided you always used the same hand as 'one hand', a 'unit' of horse height, you could compare as many horses as you wished, and rank them in precise order of height. The same technique could be used for comparing lengths of cloth, or sacks of grain, or even intervals of time, if you picked a suitable unit; each inversion of an hourglass, for instance, would indicate one more hour passed. This idea of comparison by counting units was called 'measurement'. Initially, it was a purely practical procedure, enabling traders and their customers to agree how much of a commodity was being supplied and purchased, on the basis of 'weights and measures' mutually acceptable – an 'ell' of cloth, a 'pound' of turnips, an 'acre' of land. However, in due course measurement also fostered a much more precise description of natural circumstances and events.

Scientists such as Galileo Galilei and Isaac Newton set about developing ways to compare phenomena such as movement not only with words such as 'slow' and 'fast' but also with numbers: '10 feet per second', '100 feet per second', and so on – and to represent these relationships in the symbolic language called 'mathematics'. In the following century Daniel Gabriel Fahrenheit and Anders Celsius devised ways to compare the subjective sensations of 'hot' and 'cold' with numbers, by means of 'thermometers' calibrated to give a numerical scale of 'temperature' in steps of 'degrees'. Antoine Lavoisier and Pierre Laplace invented 'calorimeters' that could measure the amount and flow of heat between hot and cold bodies, in 'calories', from the Latin word 'calor' meaning heat.

While the scientists were working out consistent and coherent ways to describe the workings of nature, practical engineers were applying their own common-sense understanding. They had known for millennia that the efforts of human and animal muscles could be enhanced by using devices such as levers, sloping ramps or 'inclined planes', ropes and pulleys and other so-called 'machines' to make 'work' easier, lifting heavy objects and moving them from place to place. Gradually the concept of 'work' emerged as another quantity that could be measured and described with units and numbers. Lifting a measured weight a measured height required so much work; lifting twice the weight the same height required twice as much work; lifting the original weight twice as high also required twice as much work.

Moreover, you could do the work by using not only animate muscles but also, for instance, a falling weight, or water running over a mill wheel, or other inanimate means. By the beginning of the nineteenth century scientists had recognized that muscles, falling weights, running water and other natural systems contained what you could think of as 'stored work', that you could use to apply forces and make things move. In 1807 the British physicist Thomas Young proposed that this stored work be called 'energy'. It was to prove a concept even more potent than Young could have imagined.

One unexpected consequence arose from the activities of Benjamin Thompson, the brilliant scientist-administrator who founded Britain's Royal Institution. In the course of his multifarious career Thompson spent a decade in the service of the Elector of Bavaria, who granted Thompson the title Count von Rumford. As head of the Bavarian War Department, Thompson supervised the manufacture of gun barrels; and he was struck by the fact that drilling out the holes in the metal castings made the metal hot. He surmised that the work done to turn the drill was related to the heat acquired by the casting; but precise measurements were too difficult to be accurate. Not until 1843, more than fifty years later, was Thompson's conjecture proved beyond doubt. In a series of elegant and scrupulously careful experiments, the British physicist James Prescott Joule demonstrated that the amount of work done by a falling weight turning paddles in a water-filled vessel was exactly proportional to the amount of heat transferred to the water.

Joule's experiments showed conclusively that work and heat were both manifestations of the same physical property: each was 'energy', and one manifestation could be transformed into the other. Joule unified what had been to that time widely disparate aspects of nature, and revealed the real potency of the concept of energy. As scientists explored its implications, they came to acknowledge energy as a fundamental attribute of the universe, a unifying principle allowing us to see the connections between the most apparently different circumstances and events.

Countless experiments, of every variety, in due course convinced scientists of a remarkable fact: energy is never created, and never destroyed. Indeed this is probably the best-known popular fact about energy – the 'law of conservation of energy'. Contrary to popular belief, the total energy consumption of the world is zero. No matter what official statistics may tell you, no one – no householder, no driver, no factory worker – 'consumes' energy. Energy is 'converted', from one form to another; but the quantity of energy always remains the same.

Energy, as we now know, manifests itself in a profusion of ways. Suppose, for instance, you shoot an arrow at a target. When you bend the bow, 'chemical energy' from the food you have eaten is converted into 'mechanical energy' of your contracting muscles. It is then stored briefly as 'elastic energy' in the bent bow and the stretched bowstring. When you release the bowstring, the stored elastic energy is converted into energy of motion, 'kinetic energy', of the arrow. When the arrow comes to rest in the target, the arrow's kinetic energy is converted into energy of deformation of the target, and by friction into heat energy; the arrow and target get slightly warmer. Some of the energy will also be converted into 'acoustic energy', the whistle of the arrow through the air and the 'thunk' as it hits the target; and so on. Throughout the whole process the total quantity of energy involved remains unaltered. This is precisely what makes energy such a useful scientific concept. By keeping track of the total quantity of energy throughout a conversion process, we can establish measurable connections between physical phenomena whose relationships seem otherwise remote or complex, like the bend in the bow and the distance the arrow travels.

We can measure the quantity of energy involved by watching the effect produced when it is converted: how much a bow can be bent, how far an arrow can be shot, and so on. However, as we have seen, until the 1840s few people realized that mechanical motion and heat were both manifestations of the same physical concept, or that one could be converted consistently into the other. People therefore came up with an assortment of 'units' for measuring energy, depending on which manifestation was being measured: 'foot-pounds', 'calories', 'British thermal units' and many, many others. The jumble of historical energy units is a grotesquely clumsy vocabulary to inflict on a concept that ought to foster simplicity and clarity. All the various energy units are convertible one into another, but the arithmetic entailed is infuriating; even energy experts get slightly shrill when discussing energy units.

We are, however, stuck with them, as we are stuck with many other awkward idioms in the language of energy; see Annex 2, 'Discussing Energy'. The energy units of the international scientific system, the Système International (SI) are gradually – very gradually – becoming the standard in most parts of the world. Fittingly enough, the basic SI unit of energy is named after the scientist who showed its universal applicability: it is called the 'joule', pronounced, some might say appropriately, considering the value of the concept, 'jewel'. To lift a bag of granulated sugar, a mass of one kilogram, its own height, about a tenth of a metre against the pull of gravity, you must convert one joule of energy. To lift the bag one metre you must convert ten joules; and so on. If you are a fifty-kilogram woman going up a flight of stairs three metres high, you must convert $50 \times 3 \times 10 = 1500$ joules of energy just to lift yourself from the bottom of the stairs to the top. As you can see, the joule is a small unit. When discussing the quantities of energy being converted by natural systems and human activities around the world, in order to have conveniently manageable numbers we use multiples of the basic unit, identified by their prefixes. Just as 1000 metres equals one kilometre, so 1000 joules equals one kilojoule; and so on, like this:

1000 joules = 1 kilojoule (kJ)
1000 kilojoules = 1 megajoule (MJ)
1000 megajoules = 1 gigajoule (GJ)
1000 gigajoules = I terajoule (TJ)
1000 terajoules = I petajoule (PJ)
1000 petajoules = 1 exajoule (EJ)

Thus, the woman climbing the stairs converts 1500 joules, or 1.5 kilojoules.

The rate at which energy is converted is called the 'power'. Converting one joule of energy per second delivers a power of one 'watt', abbreviated W; the unit of power is named after James Watt, the British scientist and engineer who provided the key to a crucially important form of power, the steam engine. Just as the joule is a small unit of energy, so the watt is a small unit of power. Human activities involve converting energy in thousands, millions and billions of watts – that kilowatts (kW), megawatts (MW) and gigawatts (GW). Even w are asleep, your body is converting energy – breathing, pumping blood, sending nerve-impulses – at a rate of about 100 watts or 100W, comparable to a bright traditional light bulb. If the woman mentioned above climbed the stairs in three seconds she would be converting 1500 joules in three seconds: that is 1500/3 = 500 joules per second, or 500 watts (500W).

Converting energy

Every change – *every* change, of anything, anywhere, any time – is accompanied by conversion of energy; but the total quantity of energy involved never changes. What changes is the net *quality* of the energy. The quality of energy depends on its organization and its concentration. The more highly organized or more concentrated the energy, the higher its quality. You can give extra energy to an arrow by firing it from a bow – organized energy, in which all the particles of the arrow move the same way; or by heating it over a flame – disorganized energy, in which the particles of the arrow move back and forth in random directions. We now know that what we have long called 'heat' is actually disorganized energy. Its concentration is called 'temperature': the higher the concentration of disorganized energy, the higher the temperature – for instance, the hotter the arrow.

As you might expect, in any energy conversion process, although the total *quantity* of energy remains unaltered, the net *quality* of the energy decreases. The kinetic energy of the flying arrow is high-quality energy because it is orderly and well organized: all the parts of the arrow move in a coherent fashion together. When the arrow embeds itself in the target, this organized energy is converted and scattered in all directions as the arrow bends and tears the target and rubs against it, making arrow and target slightly warmer. Even this small amount of locally concentrated heat energy then spreads out through the material until ever... is once again at the same temperature as the surroundings. ... high-quality energy is now disorganized and diluted to such ... t it has become imperceptible. Any energy conversion ...nilar consequences. Energy that is highly organized ...oncentrated heat energy at a high temperature spreads ...oler.

...s, however, we have a treasure house of high- ... Energy makes the world go round; and almost all the energy that does so on Earth comes from the sun. The heart of the sun pours out high-quality energy in the form of sunlight, visible and invisible, which streams into space in all directions. A minute fraction of this sunlight is intercepted, 150 million kilometres away, by the Earth. Some of the sunlight that strikes the outer atmosphere of the Earth is at once reflected back into space. Some is absorbed in the atmosphere; some reaches the surface of the Earth, directly and indirectly. The high-quality sunlight energy is converted by many different processes in the atmosphere and on the surface of the planet. Some sunlight energy is converted into heat, warming the air, the water and the soil. Some drives elegant chemical reactions in green plants, which store the sunlight energy as chemical energy in energy-rich molecules such as starches and sugars. Some sunlight energy evaporates water from rivers, lakes and oceans. As the water vapour moves through the atmosphere, sooner or later it condenses again into rain or snow, giving up the energy acquired during evaporation; you may have noticed that the air feels warmer after a rainfall. Currents in the oceans and winds in the atmosphere carry warmer water and air from place to place. Thus the sunlight energy that reaches the Earth during any given hour is soon apportioned all over the planet, as ambient energy, the energy of the local environment. This ambient energy in your surroundings is converted continually from one form to another, its net quality decreasing as it drives the weather, the growth and activities of plants and animals, and all the other physical, chemical and biological processes taking place everywhere.

However, the Earth does not acquire this energy to keep. Energy tends to move from a warm place, where the heat energy is concentrated, to a cooler place, where the heat energy is more dilute. In outer space the heat energy is very dilute indeed; the temperature goes down to about 270 degrees below zero Celsius, only 3 degrees above absolute zero. Needless to say the Earth is considerably warmer than this; so the Earth, too, sends a stream of energy – 'Earthlight' – in all directions into space. The apparent temperature of the Earth, when viewed from space, is determined by the balance between the high-quality sunlight arriving on the Earth and the low-quality 'Earthlight', invisible infrared light, leaving the Earth. The Earth's 'average' temperature is such that the energy arriving is offset by the energy leaving, so that the 'average' temperature remains the same. If energy were accumulating on the Earth, its 'average' temperature would rise until the amount of energy leaving was again in balance with the amount arriving. Of course the Earth's 'average' temperature when viewed from space is an average over a wide range of prevailing temperatures in various parts of the Earth's atmosphere and on its surface, from more than 50 degrees below zero Celsius to more than 50 degrees above. The whole system is extraordinarily complex, and as yet far from well understood. But its behaviour has far-reaching implications for our future attitude to energy, as we now understand all too clearly.

More than 99 per cent of the energy conversion that takes place on Earth involves sunlight energy in one form or another. The quantity of this energy arriving and leaving is enormous; and almost all the conversion it undergoes takes place willy-nilly, with no intervention by human beings. The motions of the atmosphere and the oceans, the progression of the seasons, the variations of outdoor temperatures from place to place, from hour to hour and from year to year, the life and death of wild plants and animals, all are driven by converting the ambient energy of sunlight.

We humans, too, rely on sunlight for almost all the energy conversion processes of importance to us. Sunlight keeps the average temperature of the Earth almost high enough for human bodies to survive and function as living organisms with no additional assistance; look at any popular beach resort in summer. Sunlight energy stored in plants provides food for people, either directly when we eat the plants or indirectly when we eat animals that have eaten the plants. People, including energy experts, tend to call energy 'useful' only when they notice that they are using it; but none of us would survive without the continuing support of the natural ambient-energy conversion processes we take for granted. The energy we 'use' and call 'useful', because we intervene consciously in its conversion, is a modest, not to say trifling, addition to the natural ambient-energy conversion processes on which we unconsciously rely.

Portable energy

Nevertheless, the human use of energy differs in one unique and telling respect from the use of energy by all other living things. Human beings have learned to use 'fuel'. As noted earlier, fuel means 'material for a fireplace': material whose energy content can be mobilized under human control, and converted at places, times and rates chosen by humans. However, our unique ability as humans lies not merely in recognizing the concept of fuel as a store of accessible energy. The significance of the myth of Prometheus, the human acquisition of fire, was not the mere discovery of the process of burning; fire occurs in nature, as a result of lightning, spontaneous combustion and other causes. The significance of the myth was the discovery that a human could *control* fire: that is, a human could start a fire at a chosen place and time. No other living creature has mastered this skill. Other living creatures grow and shed fur or other body coverings, to retard or accelerate the loss of heat from their bodies, and seek or build shelters to provide comparatively stable temperatures around them. But no other creature can mobilize at will energy stored *outside* its body, to initiate a controlled process of energy conversion, and raise the local temperature to a level much higher than the ambient temperature. Of all the characteristics that distinguish humans from other living creatures, the ability to use fuel may be paramount, possibly even more distinctive than our development of language. Not even the great apes use fuel.

No one will ever know just how humans learned to use fuel. Perhaps, one day, one of our ancestors found a tree burning after a lightning strike, and liked the warmth and light it provided. Our ancestor might have noticed that the fire could spread from one branch to another; perhaps, as the fire began to die, our ancestor took the initiative and placed another dry branch nearby, to see if it too would begin to give off warmth and light. It seems likely that for a long time humans could only feed and cherish fires started by natural occurrences; actually starting a fire must have been accomplished much later. Even today you will have great difficulty starting a fire using only natural materials, if you do not resort to a sophisticated human device like a match or a magnifying glass. If you try, you will find that you need dry fuel, preferably finely divided and of low density, for instance dry leaves. You must heat this fuel until it is much hotter than ambient temperature, when the fuel will begin to react rapidly with the oxygen of the air. This reaction, once started, gives off more heat, until the reaction is occurring throughout a significant volume of the fuel: it has 'ignited' and started to burn. Your problem is to bring about the initial temperature rise, to start the reaction.

If you have only natural materials at ambient temperature, you have two possible ways to raise the temperature of a material until it ignites.

You can take two suitable rocks and bang them together, hoping that the mechanical energy of your muscles will be concentrated into an abrupt and violent tiny fracture at the point of contact of the rocks, producing a speck of very hot broken rock called a spark. If this spark lands in the finely divided fuel, say the dry leaves, the spark may transfer its heat to the fuel and raise its temperature to ignition. Of course you may have to spend a long time banging the rocks together before you get a hot enough spark to land just right. Alternatively, you can take two pieces of wood, suitably shaped, and rub them vigorously together so that the mechanical energy from your muscular motion will be converted into frictional heat at the point of contact between the pieces of wood. If you place your finely divided fuel at this point you may be able to raise its temperature enough to ignite it. Of course your arms may get very tired in the process. In either case the concentrated high-temperature heat you have laboriously generated by your efforts may leak away into the surrounding air too fast, so that the fuel never gets hot enough to ignite. In retrospect, the human achievement of learning to start a fire is very impressive indeed, as you can confirm for yourself the next time you take a walk in the woods.

Nevertheless our ancestors did learn to start a fire. In doing so they acquired an ability whose consequences have been incalculable. The primary result of this ability is that a human can create, at will, a local temperature far higher than that found normally in the parts of nature accessible to humans. Access to warmth and light independent of sunlight has allowed people to survive even in polar winters, and colonize almost the entire planet. Furthermore, such high temperatures foster many interesting phenomena. We can cook food that would be inedible or unhealthy uncooked. We can harden and glaze earthenware vessels to make them watertight and heat-resistant. We can smelt ores to yield metals, and shape the metals into tools. Worked metal tools in turn make it easier to prepare fuels and ignite them. In short, the ability to start a fire made it thereafter progressively easier to start a fire.

The ability to use fuel also allowed people to stockpile a store of energy that could be tapped as desired. They could carry this stored energy from place to place and use it when and where they wished. Fuel was thus not only an energy store but also an energy carrier. Throughout the prehistoric eons humans discovered a wide variety of materials that would burn, including wood and other dried plant materials, dried animal dung, animal and vegetable fats and oils, peat and coal. They also built increasingly elaborate structures and devices to utilize the heat from fuel: ovens, kilns, furnaces, braziers and so on. With tools made from metals recovered and shaped by using fuels, people also created structures and devices that could control for human purposes the conversion of natural ambient energy carried by falling water and blowing wind: water mills and pumps, windmills and sails.

Organizing energy

However, despite the rapidly expanding opportunities for using fuel, one major limitation remained. Energy converted by burning fuel manifests itself as heat. Heat, no matter how hot, is disorganized energy. The energy of material motion, for instance that of a flying arrow, is organized. Until less than three hundred years ago there was no practical way to convert disorganized heat energy from fuel into organized energy of motion, like that of machines, 'mechanical' energy.

Accordingly, although people could use fuel-energy for many other purposes, they could not convert it into mechanical energy. If they wanted to move anything, they could use only food energy or ambient energy – the muscles of people and animals, or the structures of mills and sails. The strength of muscles is limited; water energy could be used only where the water was; wind energy was variable and unpredictable. However, by the end of the seventeenth century an idea was taking shape that eventually changed the face of the world. It was called the steam engine. In the steam engine, you use fuel energy to boil water, turning it into steam and increasing its volume a thousandfold. You then condense the steam back to water, with a corresponding decrease of volume. The changes of volume of this 'working fluid' move a piston back and forth in a cylinder, converting some of the disorganized heat energy from the fuel into organized mechanical energy. The steam engine thus overcame the crucial limitation on human use of fuel. By turning fuel energy into mechanical energy, the steam engine changed the course of human history.

Disturbing energy

The coal-burning steam engine became a crucial factor in the 'Industrial Revolution', which dramatically transformed production methods, population patterns and social organization. Many commentators have offered historical evaluations of the Industrial Revolution, its social, political and ethical debits and credits. One consequence of the Industrial Revolution, however, is belatedly attracting urgent attention. Fuel use not only increased steadily but shifted from firewood, animal dung, tallow and the like, to coal, and thereafter to petroleum and natural gas. Firewood and other fuels derived directly from animals and plants contain sunlight energy that has been stored in the material for at most a century or so. When such fuels are burned they release this stored sunlight energy again, returning it to the Earth's global systems from which the energy will duly be reradiated into space. The entire process of storing sunlight energy, burning the fuel to release it again, and reradiating it into space as 'Earthlight' takes place over a reasonably brief period, and keeps the Earth's 'energy budget' in balance. Storing sunlight energy in green plants also entails extracting carbon dioxide from the atmosphere, and incorporating the carbon into solid plant material such as starches and sugars. When the plant material is burned, the stored carbon is turned again into carbon dioxide and returned to the atmosphere, thus also balancing the Earth's carbon dioxide budget.

Coal, petroleum and natural gas, by contrast, are fossil fuels. Coal is the mineral or 'fossilized' remains of jungle vegetation that flourished on Earth more than 150 million years ago. Petroleum is the fossilized remains of small marine creatures that thrived in the oceans at about the same time. Natural gas was formed by the decaying vegetable and animal matter as it was turning into coal and petroleum. Accordingly, the energy contained in these fossil fuels arrived on Earth as sunlight not decades but hundreds of millions of years ago. Since its arrival the average temperature of the Earth has undergone a number of major swings, and even created several ice ages. Burning fossil fuels, and releasing this 'fossil sunlight' hundreds of millions of years after its arrival on Earth, introduced a new factor into the energy balance of the planet, a time lag on a scale far longer than any associated with firewood and similar fuels.

From the mid-1940s onwards, the development of nuclear energy introduced a further complication. Nuclear energy is released from the innermost parts of the atoms of certain metals, in particular uranium. Uranium is thus another type of fuel, a 'nuclear fuel'. Nuclear energy, like fossil sunlight from fossil fuels, is out of balance with the Earth's current energy budget of sunlight arriving and Earthlight leaving.

Indeed nuclear energy is in a category of its own: unlike fossil sunlight, nuclear energy has never previously passed through the Earth's physical or biological systems.

Until the significant use of fossil fuels, the average temperature of the Earth had been established for millennia by the balance between sunlight arriving and 'Earthlight' leaving, as mentioned earlier. However, the mobilization of fossil sunlight in increasing quantity has for more than a century been injecting a gradually growing amount of energy into the Earth's conversion processes, in addition to the energy arriving daily from the sun. Averaged over the entire Earth, to be sure, the fossil sunlight contribution is still minute, and the nuclear contribution much smaller still. However, local side-effects of fuel use are long since all too familiar: 'heat islands' over major cities regularly create anomalous weather phenomena, and aggravate air-quality problems caused by noxious products from burning fuels.

Local side-effects are bad enough. But we now know beyond reasonable doubt that using fossil fuels also has global side-effects; indeed to call them 'side-effects' is a drastic understatement. Burning any fossil fuel releases its fossil carbon, carbon extracted from the atmosphere and stored in solid form many millions of years ago, back into the atmosphere as carbon dioxide. This fossil carbon dioxide is an addition to that which has in recent millennia been in equilibrium with the Earth's plants, animals and natural systems. Since the 1850s the average concentration of carbon dioxide in the Earth's atmosphere has increased by more than 25 per cent. Carbon dioxide is not of course toxic: on the contrary, it is an essential ingredient of life. Plants inhale carbon dioxide whenever sunlight is shining on their leaves, and exhale oxygen. But carbon dioxide has another attribute, one that has made it the world's most controversial gas.

Heating the greenhouse

In the atmosphere, carbon dioxide molecules absorb some of the 'Earthlight' heat energy leaving the Earth's surface; and instead of reradiating it onward into space they radiate a fraction of it back to the Earth. Atmospheric carbon dioxide therefore acts rather like the glass in a greenhouse. Greenhouse glass transmits high-quality visible light energy essentially unimpeded; you can see through the glass. Once inside the greenhouse, however, this high-quality energy is degraded to low-quality heat energy. The heat tries to emerge again from the greenhouse as invisible infrared radiation; but the glass reflects some of the heat radiation back into the greenhouse, raising the temperature inside. In the same way, atmospheric carbon dioxide reflects some of the escaping heat energy back to the Earth, and thereby raises the average temperature at the Earth's surface.

Other gases in the atmosphere do the same, among them water vapour, methane, nitrous oxide and the 'chlorofluorocarbons' or CFCs, once best known as propellants in traditional aerosol spray cans. CFCs are the most effective 'greenhouse gases', 20,000 to 30,000 times as effective as carbon dioxide, molecule for molecule; and the other 'greenhouse gases' are also contributing to the overall change in the Earth's heat balance. Nevertheless, atmospheric scientists are now convinced that carbon dioxide is responsible for well over half the total greenhouse effect. Measuring the consequent temperature rise is difficult, not least because temperatures already vary so much from place to place and through the year. But responsible scientists around the world now concur that 'global warming' by the greenhouse effect is already underway. Some estimates suggest that the average temperature of the Earth may rise more than 5 degrees Celsius in this century. Such a rapid temperature rise would have a profound and potentially catastrophic impact on climate and weather, water supplies, the habitat of wild plants and animals, agriculture, coastal zones, major cities and indeed every aspect of life on Earth.

Scientists are urgently studying the build-up of atmospheric carbon dioxide to ascertain its causes and its effects. Burning vast tracts of tropical rainforest not only turns the tree-carbon into carbon dioxide, but removes the green leaves that would inhale the carbon dioxide once again. Marine pollution such as oil slicks keeps submicroscopic marine plants called 'phytoplankton' from absorbing carbon dioxide from the atmosphere above the oceans. Nevertheless the observed increase in atmospheric carbon dioxide within the past century has undoubtedly arisen largely because humans have mobilized fossil carbon by burning fossil fuels. The more fossil carbon we inject into the Earth's carbon budget, the more we shift the balance towards a higher atmospheric concentration of carbon dioxide. The Earth's climate depends on systems of bewildering complexity, sensitive to even slight disturbance. If we upset these systems, no one can say with confidence what will happen, or how fast. We are conducting a vast collective global experiment, whose result is impossible to foresee and impossible to reverse.

As we noted in Chapter 2, official pronouncements about the future use of fuels continue to presume that such use will increase substantially in the years to come. Until recently such presumptions raised doubts only about the cost and availability of the fuels and the technologies that require them. Now, however, scientists, politicians and other concerned people everywhere recognize that the human use of fuels has global side-effects that grow ever more alarming. We have begun to realize, all but too late, that the Earth cannot tolerate unlimited human intervention in natural systems, including energy systems. Within the present generation, we fuel-using animals have to come to terms with limits on the use of fuels. If the Earth is to continue to tolerate human life, we have to find a way to make energy use sustainable.

We can start by transforming electricity.

Part II

MAKING ELECTRICITY SUSTAINABLE

4

FULL CIRCLE

After more than 120 years, electricity may be starting to come full circle.

In the 1870s, the first systems for electric light – the generator with its motive power, the cables, the switches and the lamps – were all together on the same premises. The owner, say, of a stately home desiring the striking new status symbol of electric light had to buy, operate and control the entire system. The light, from arc-lamps, was noisy, smelly and inconvenient; but it was spectacular. It was also, of course, ostentatiously expensive; the cost alone made it a status symbol, not available to the common herd. The owner-operator was buying, and paying for, electric light – not electricity.

Thomas Edison's great idea was simply to scale up the whole process, to reduce its unit costs. The motive power – steam engine or water wheel – and the generator attached to it both exhibited substantial economies of scale. A steam engine or water wheel ten, twenty or fifty times the size would light ten, twenty or fifty times as many lamps, without costing ten, twenty or fifty times as much. Indeed the larger the generator, the lower the cost per lamp, even allowing for the extra cost of more and longer cables. A system so large would light more lamps than the most extravagant single individual could possibly desire. To demonstrate the effectiveness of this arrangement, Edison therefore had to recruit a roster of different property-owners on whose separate premises he could install lamps, all connected to the same central generator.

Even with the largest steam engine then feasible, the overall conversion of coal energy into electricity was well under 10 per cent. Edison's incandescent lamp in turn converted less than 10 per cent of the electricity into light. At least 99 per cent of the coal energy was therefore lost as waste heat. In order to keep the cost from being even more prohibitive, Edison had to minimize waste, and optimize the entire system –

This chapter is adapted from 'Full circle', *Cogeneration and On-Site Power Production*, January–February 2000.

engine, generator, cables and lamps – to make it as efficient as possible. His first such systems, in Holborn, London, and in lower Manhattan, went into operation in 1882. The Manhattan station, with a single large steam engine and generator in a building on Pearl Street, lighted offices on Wall Street and at the *New York Times*; Edison always knew how to attract attention. Edison billed his customers for the light they used, according to the number of lamps on each premises.

Then, in the mid-1880s, it all began to go wrong. No one realized this at the time. Even today the profound importance of the crucial misstep is still lost on almost all participants. But a bald statement of the facts leaves no room for doubt. The single most effective deterrent to improving the efficiency of electricity systems is the electricity meter. From the moment it was introduced, shortly after the start-up of the Pearl Street system, the electricity meter changed the ground rules. Its pernicious effects have been felt ever since. If what you are selling is electric light, you want the whole system to be as efficient as possible. If, on the other hand, you are selling units of electricity as measured by an electricity meter, someone using less efficient lamps has to buy more electricity from you to get the same level of illumination. From the point of view of you, the seller, inefficiency on the customer's premises is good for your business.

The received wisdom, of course, sees the matter very differently. It considers electricity to be a commodity like natural gas or water, delivered to your premises for you to use as you wish. The meter just measures the flow of the commodity; the supplier bills you accordingly. Early in the twentieth century steam engines with reciprocating pistons gave way to rotating steam turbines, larger and larger, until the fuel efficiency of such 'steam-cycle' units reached 30 per cent or even more; the largest machines even passed 40 per cent. The economies of scale of ever-larger steam-turbine and water-turbine generators steadily reduced the cost of a unit of electricity, so much so that electricity is now ubiquitous in modern industrial society, indeed taken completely for granted. Since the beginning of the 1990s, liberalization and the introduction of competition have underlined the view that electricity is a commodity. The whole market apparatus laboriously erected across Europe, North America and elsewhere is based on this presumption.

Unfortunately, however, electricity is not a commodity. If you are selling a commodity, you can store it and hold it back from the market until you get the price you want. If, however, you are selling electricity, you cannot store it to sell later for a better price. Nor, despite frequent usage to the contrary, is electricity a fuel. As we noted earlier, a fuel such as coal, oil or natural gas is a physical substance. It comes out of a hole in the ground at a particular place. If you want to use it anywhere else you must physically transport it there. Electricity, by contrast, is a physical phenomenon happening instantaneously throughout the entire interconnected system, including all the end-use equipment. Moreover, you can generate electricity anywhere, in many different ways.

However, the traditional configuration of electricity system uses large-scale remote central stations to generate electricity as synchronized alternating current, delivering it to users over a network of cables usually including substantial lengths of high-voltage transmission lines. This configuration, the common technical model replicated all over the world in the past half-century, arose for Edison's reason: scaling up the generating technology powered by steam or water turbines reduced the cost of a unit of electricity. Other technologies for generating electricity have long been available, everything from diesel generators to wristwatch batteries. But the electricity they produce costs more per unit, often a great deal more; imagine powering even a single 60-watt lamp with wristwatch batteries.

Generation innovation

In modern industrial countries the remarkable success of the traditional electricity system into the 1980s confirmed and reinforced the underlying tacit view of electricity as a commodity. This in turned enabled free-market theorists to launch the process of liberalization of electricity. It began in Chile and the UK in the late 1980s, and spread thereafter at an accelerating rate over much of the world. By a remarkable coincidence, liberalization got under way just as a new fuel, natural gas, was emerging as a serious option for electricity generation in many parts of the world. Moreover, you could use this new fuel in generating technologies whose economies of scale were very different from those of traditional steam-turbine and water-turbine generators.

The first breakthrough technology was the gas turbine. A gas turbine burns premium fuel, making very hot combustion gas that expands rapidly and turns the rotating shaft directly. What we call 'jet engines' are gas turbines, shooting a stream of this hot combustion gas out of the rear nozzle at velocities exceeding the speed of sound. Gas turbines have powered aircraft since the 1940s. By the 1980s, technology development for aircraft engines had made gas turbines sufficiently powerful, reliable and cheap that you could use one to turn an electricity generator in continuous operation. A gas-turbine generator can be efficient and economic at a much smaller size. You can order and install it and have it in operation in under two years. Firing natural gas, it requires no fuel storage. It produces no solid waste, and its emissions can be very low. You can therefore site it much more easily, close to users and indeed on the site where they want to use the electricity. The exhaust combustion gas from a gas turbine is so hot that it can even be used to raise steam for a steam turbine. Operating gas and steam turbines together in so-called 'combined cycles' substantially increases the total electricity output from a given amount of fuel. The overall fuel efficiency for the most modern large combined-cycle stations can be 60 per cent, compared to well under 50 per cent from the most modern pure steam-cycle units. Gas-turbine generation also lends itself well to so-called 'cogeneration' or 'combined heat and power', producing both electricity and usable heat, with overall fuel efficiency which can be well over 80 per cent.

In the early 1990s, in the first rush of enthusiasm for liberalization, new gas-turbine stations were usually aggregations of generators on a single remote site, essentially equivalent to traditional steam-turbine and water-turbine stations in the centralized configuration of the prevailing traditional system. Gradually, however, some electricity people noticed that gas-turbine technology makes smaller stations closer to users not only feasible but frequently desirable, reducing the need for long transmission lines and the accompanying losses, especially when you can locate generators actually on site. Siting more and smaller generators closer to users is a sharp break with the traditional trend towards ever-larger stations ever farther away.

At the same time other generating technologies began to emerge for serious consideration (see also the glossary, pages 177–184). Conventional petrol and diesel internal combustion engines have long been used for free-standing intermittent or emergency generators of modest size, with no network connection. But they require expensive fuel, and their noise and fumes make them unwelcome neighbours. From the 1990s, however, a first cousin, the internal combustion gas engine, burning the newly cheap and abundant natural gas, has become a popular and successful alternative in many places, especially for on-site cogeneration of electricity and heat. As with larger cogenerators, the hot exhaust gas from the gas engine can produce hot water in quantity, for direct use or for central heating, just like an ordinary hot-water boiler, but one that produces electricity as a bonus from the same amount of fuel. The robust and familiar technology operates automatically with simple controls, needing no supervision or nearby staff.

An even older technology, the Stirling or 'external combustion' engine, offers a similar option on an even smaller scale, down to a kilowatt of electricity plus hot water, a size suitable for an individual household. Several major suppliers are now bringing to market Stirling-engine so-called 'microcogen' units for new houses, and as replacements for old central-heating boilers, in the UK and elsewhere.

Microturbines, the size of a wardrobe, with a single moving part, the high-speed shaft, can burn natural gas to generate perhaps 25kW of electricity. Microturbines can also be used in cogeneration mode, boosting overall fuel efficiency by delivering useful heat or hot water as well as electricity.

The oldest fuel of all, firewood, and all the other forms of combustible 'biomass', can now deliver heat and hot water more cleanly, conveniently and efficiently, in compact modern boilers burning wood chips and other processed biomass. But a yet more attractive possibility is biomass gasification: roasting the fuel to produce a combustible gas to use in gas-engine generation or cogeneration – potentially a major technical option for parts of the world with abundant biomass and no central electricity. Such biomass power technology is still under development; but the pay-off could be vast.

One further option for fuel-based electricity generation is perhaps the most appealing of all. The concept of the 'fuel cell' dates back to 1839, when the British scientist William Grove first proposed it. In a fuel cell, hydrogen reacts directly with oxygen from the air to form water and produce an electric current between terminals like those of a battery. Because the fuel cell uses a continuous supply of hydrogen it is sometimes called a 'flow battery'. Many different kinds of fuel cell have been developed and demonstrated, with outputs up to a megawatt or more, and down to less than a milliwatt, suitable for a laptop or a mobile phone. Fuel cells work; they were already in use in space vehicles in the 1960s. But two factors still hamper their wide-scale application. One is the need for hydrogen, or a suitable process to produce it in or near the fuel cell. The other is the need to bring down the cost. Until manufacturers tool up for mass production, fuel cells are essentially handmade, one by one, and accordingly costly. But until manufacturers can see a large enough market for them, they are reluctant to invest in facilities for mass production. If and when fuel cells at last come into their own, one further refinement awaits – the biomass gasification fuel cell, which could be the ultimate village-scale generating technology in rural areas.

The fastest growing of the innovative generating technologies worldwide is wind power. Modern wind turbines are far removed from the 'windmills' once common across Europe and elsewhere. Blade materials and aerodynamics, power train improvements and controls have combined to make machines able to generate upwards of two megawatts per unit reliably and with minimal maintenance. 'Wind farms' both on land and offshore are now making a substantial contribution to electricity systems in Denmark, Germany, Spain, India, the UK, China, the US and a lengthening list of other countries. The biggest problem now facing prospective wind-power developers is access to turbines. Manufacturers have order books filled for many months ahead. Meanwhile, a recent upsurge of interest has created a market for much smaller wind generators, suitable to install on individual buildings and even houses. The economics of these small units are still controversial, but in some locations at least they can clearly make a useful addition to the generating mix, especially in rural areas with good wind conditions.

Wind, of course, has delivered useful energy for millennia; so has water. Traditional hydroelectricity has been a major contributor to electricity systems since they were first established. Like steam power, water power was scaled up to enormous dimensions, with massive dams flooding vast areas; but it does not have to be. Some smaller hydro installations divert modest river flows through simple turbines to generate up to a few megawatts with minimal disturbance. Others sit in the river flow itself, passing some flow through turbines in 'run-of-river' generators. Such 'minihydro' and 'microhydro' units are especially suitable for local systems supplying local loads. In Nepal alone they number in many hundreds.

Marine energy, in waves and marine currents, is tempting, both concentrated and moderately predictable. But the offshore environment is challenging even without storm conditions, making installation and maintenance a severe constraint thus far. Development work nevertheless continues in many places, because the prize is eminently worth the effort. Tidal energy has vociferous enthusiasts; but the opportunities are limited, and would have to avoid deleterious effects on wetlands and estuaries.

Solar hot-water heating is long since well established in suitable latitudes, and its range could well be significantly extended. So-called 'solar thermal' electricity could likewise become a significant technology, concentrating the sun's rays to produce steam for a turbine; units are already in operation in California and elsewhere. But probably the single most appealing form of innovative generation is direct solar electricity, or 'photovoltaics'. Turning the energy of sunlight directly into electricity is already a practical reality, widespread and rapidly expanding. So-called 'solar panels' now include not only free-standing panels but roof tiles, window glass and other cladding for buildings, which become part of the structure – a part that generates electricity. At the moment the various forms of technology for solar electricity generation are usually considered expensive; but costs are coming down steadily. In any case such cost comparisons are invariably skewed by asssumptions carried over from traditional electricity, disregarding risks and environmental impacts.

Innovation implications

Since the mid-1990s the possibility of incorporating such small-scale generation and of decentralizing electricity systems has been looking both increasingly feasible and increasingly attractive. As yet these small-scale technologies often appear more costly than traditional options considered in the traditional context. But liberalization is also changing the financial ground rules. In a traditional monopoly franchise, captive customers guaranteed a revenue stream to support large-scale long-term projects such as gigawatt-scale power stations and long high-voltage transmission lines. In a liberal context such projects become acutely risky, not for captive customers but for company shareholders and bankers. The new financial ground rules have already affected the choice of electricity technologies. The effect will intensify as other small-scale options prove themselves.

The consequences may nevertheless prove uncomfortable. Traditional monopoly systems usually include substantial redundancy – extra generating and network capacity, available for use in case of faults and failures, not to mention staff to cope, all paid for by captive customers. In a liberal context both redundancy and staffing levels drop, sometimes dramatically. At the same time, major electricity users with significant heat loads find the option of on-site cogeneration increasingly attractive, since they no longer have to face punitive charges from a monopoly for backup connection to the network. Major users, however, tend to represent the large stable loads that are easiest for the large inflexible generators of a traditional system to supply. As these loads migrate to cogeneration and leave the system, the remaining load profile on the system grows peakier, more difficult and expensive to supply. System reliability and power quality may deteriorate. If they do, more and more users may resort to on-site power generation, whose reliability and power quality the user controls. Even if on-site generation is more expensive, the extra cost may be justified as a form of insurance against the failure or poor quality of supply from the network.

Reliability and control may prove to be potent drivers of the move towards on-site generation. As microcogeneration, microturbines, fuel cells and other small-scale generating technologies mature, more and more places with ever-smaller loads will become candidates for on-site generation – not only industrial sites but office buildings, shopping malls, airports, railway stations, hotels, hospitals, schools, blocks of flats and perhaps even individual residences. What this will do to the rest of the electricity system over time is still an open question; but it may be progressively disruptive. In due course it may even put those without access to on-site generation at a severe disadvantage, a corollary as yet inadequately considered. With an abundance of options to choose from, major players will be able to take care of themselves. But who will ensure that poor neighbourhoods and rural areas still have access to electricity services? Will industrial countries, like too many developing countries, divide into electricity 'haves' and 'have-nots'? No matter who owns a liberalized system, if the lights start going off, the government will be in the front line.

Bypassing the meter

A much more positive possibility, however, also arises. After more than a century, the re-emergence of on-site generation brings with it the promise of overcoming at last the pernicious effect of the electricity meter. If you generate your own electricity on site, no one benefits by having you use inefficient buildings and equipment. Instead, like Edison on Pearl Street but with technical options that would astonish him, you can seek to optimize the whole local system. Nor must you do it yourself. In a liberal context, electricity companies are already learning that competing to sell anonymous units of electricity at a customer's meter is a precarious business. They can compete only on price; their margins become vanishingly small. If, at the same time, customers can switch suppliers more or less at will, this form of business is a good way to go bankrupt. Enlightened companies are already seeking different ways to win customers and retain their loyalty. After years of frustration the age of the genuine energy service company may be dawning at last.

Local electricity systems with on-site generation may prove a potent manifestation of the new business now emerging. In your own economic interest you and your energy service company will want to ensure that your buildings, lighting, motors, and electronics use your own electricity as efficiently as possible. Optimizing the whole local system makes economic sense; and economics and environment point in the same direction. Using less fuel to get better services is a crucial first step towards sustainable electricity.

How this will work out in practice no one yet knows; and it won't happen overnight. But after more than a century electricity may eventually come full circle, back to where it belongs: on site.

5

THE ELECTRIC CHALLENGE

Electricity is different. Those still preoccupied with electricity's many immediate problems worldwide have yet to appreciate the profound and far-reaching long-term implications of the electric difference. Over time the distinctive difference of electricity is going to alter fundamentally the way we think about energy, the way we use energy, and the way we pay for it.

For more than a century we have treated electricity as though it were a fuel. We have regarded electricity as a substance like coal or oil, as a commodity, bought and sold by the unit. In the early 1990s, when governments were liberalizing their electricity systems, they proceeded accordingly. They set up electricity 'markets' all over the world to trade in electricity as a quasi-commodity. Buyers and sellers, governments and regulators, focus their attention on the price of a unit of electricity. But electricity liberalization coincided with a wave of technical innovation that is continuing, and gaining momentum. Innovative electricity technologies are already beginning to affect the structure and function of electricity systems in many places. As the process continues, it will gradually alter the role and nature of electricity in human energy systems. In so doing, it may also reveal new ways to tackle crucial global problems, including climate change. But the transition may be disturbing and disruptive. Even as electricity policy grapples with pressing short-term issues, it must prepare to meet a challenge more severe than ever before.

As we noted earlier, traditional electricity is based on a common technical model now a century old, replicated all over the world for more than half a century. In this model large, remotely sited central stations generate electricity and send it out as synchronized alternating current, over a network including long high-voltage transmission lines.

This chapter is adapted from 'Overview: The electric challenge', Working Paper 1 for *Keeping The Lights On*, Chatham House, 2003.

The relevant government grants the system a monopoly franchise; in the franchise area no one else may generate electricity to sell. However, traditional electricity based on this model is already in trouble. Its key technologies – large dams, large coal-fired and nuclear stations, and overhead transmission lines – all face financial and environmental problems that may become insuperable. In any case, traditional electricity has failed to reach a third of the people on Earth. As traditional electricity struggles with its mounting problems, an innovative alternative is now emerging. As yet the manifestations of this electric alternative are limited; but its potential impact is extraordinary.

Governments liberalized their electricity systems by restructuring, selling off the assets to private investors, and introducing competition between generators and suppliers of electricity. Governments nevertheless appeared to believe that nothing else would change – that the systems would continue to look much the same, and function technically in much the same way. But introducing competition shifted major risks away from the captive customers of the monopoly onto the shareholders and bankers of the liberalized system. On systems thus far liberalized, the market deals in transactions in ephemeral units of electricity. An asset to produce or deliver electricity earns revenue only when it is functioning and participating both instantaneously and continuously in the process. If you can't sell its output you can lose a lot of money very fast, as many generators have recently found. Accordingly, in a competitive framework, traditional large-scale generating stations, long-term investments, become acutely risky.

Electricity liberalization also happened to coincide with the emergence of gas turbines for continuous electricity generation, and cheap and abundant natural gas to fuel them. You could order, commission and bring into service gas-turbine generation, initially in the form of combined-cycle stations, much more rapidly than traditional generation. You might take at least six years to plan, build and commission a traditional power station – often much more. However, you can have a gas-turbine station producing both electricity and revenue in perhaps two years or less. Gas-turbine generation, less risky, cheaper, cleaner and more convenient than traditional generation, is now the technology of choice for new generation wherever electricity is liberalized and natural gas available.

Liberalization has therefore triggered a shift in choice of generating technologies. One advantage of electricity is that you can generate it in many different ways, at an astonishing range of different scales. In traditional electricity, a better power station was always a bigger power station farther away. Since liberalization, however, starting with gas-turbine generation, a better power station is more likely to be smaller and closer to users – possibly even directly on the site where you want to use the electricity. As we noted in the previous chapter, other smaller-scale generating technologies are now emerging, including micro cogeneration, microturbines, fuel cells, microhydro, wind turbines of many sizes, biomass gasification and photovoltaics. Innovation in generation, however, has outstripped innovation in networks. A traditional network is radial and one-way, designed to carry large flows of electricity in one direction from large remotely sited power stations, to subdivide the electricity and deliver it to loads many thousands or even millions of times smaller. Network planners never intended it to serve as the basis for a competitive market in wholesale electricity. Nor did they expect it to accommodate large numbers of generators in sizes broadly similar to loads, probably connected to the network at voltages much lower than traditional transmission voltage. The consequent stresses and conflicts, over access to and use of electricity networks, are already evident. They will get worse.

Market and mismatch

Analysts and commentators now make much of the newly instituted 'market' in electricity, as though it were equivalent, say, to the long-established worldwide market in oil. They say less, however, about the role of regulation in determining the unit price of electricity, even in this so-called 'market'. Regulators lay down the financial rules to manage and use electricity networks, which they still almost always treat as monopolies. Even in a liberalized context such as that of northern Europe, a substantial part of the cost of a unit of electricity is the charge for use of the network. This charge has only a tenuous connection with any 'market'. Rules for using the network are essentially arbitrary; a regulator imposes them unilaterally, invoking powers the relevant government has given it. Only within this arbitrary framework does any residual 'market' exist. The regulator has a profound influence also on maintaining, expanding and altering the network. That in turn affects how the network interacts with generation and loads. Until recently the regulators of liberalized electricity systems have shown little interest in having networks evolve away from their traditional configuration and their traditional role. Such evolution is still at best preliminary and tentative. Until and unless networks evolve from a radial one-way towards a meshed two-way configuration, they will constrain severely the opportunities for fundamental change associated with small-scale decentralized and local generation. But the pressure for change is steadily intensifying.

One option attracting increasing attention is that of so-called 'private wires', to link local generation with local loads that may not be on the same site as the generation but are not too far away from it. The local generator or generators, perhaps jointly with the local loads, own and operate such private wires, which are separate and distinct from the regulated network. They thus effectively undermine its monopoly, at least in this local area. At the moment, however, legislation and regulation usually prevent private-wire participants from using the regulated network. The 'either-or' nature of the private-wire option as presently available limits its attraction, and probably entails less than optimal use of the assets involved. Private wires nevertheless point the way to the much more radical reorganization of the structure and function of networks that will almost inevitably ensue.

The stubborn and lingering mismatch between innovative generation and traditional networks, both technically and institutionally, demonstrates the pervasive inertia of what in the US are called legacy assets and legacy institutions. More stultifying still is the legacy mindset – traditional concepts and ways of thinking that continue to shape our expectations about the role and nature of electricity in society. Lifting the burden of this legacy mindset will be a major challenge to policy analysts and policymakers.

Why meter?

Traditional electricity people think of the network essentially as a delivery conduit, to carry measured units of electricity from generator to load. Thinking of the network this way severely limits the role and function of generators, particularly small decentralized generators. Remember, however, that electricity is different. For electricity in any context, whether digital watch or aluminium smelter, the network is not just an ancillary delivery system. It is an integral and essential part of the electricity process. We ought to emphasize that expression, the 'electricity process'. It underlines the change of mindset, the changing conceptual framework, we now need to invoke. Electricity as we use it is not a quantity, much less a substance or commodity. Electricity is a process. We direct the process to achieve a remarkable variety of objectives. Our objectives, such as comfort, illumination, motive power, refrigeration and information handling, we can call 'electricity services'; but electricity by itself is useless. It can deliver the desired services only by activating appropriate technology.

For historical reasons, we have come to focus on the measured amount of electricity that flows through the process and the technology. Historically, participants on electricity systems have been divided into two separate groups, one considered as suppliers and the other as users. In transactions between suppliers and users we have considered electricity as equivalent to a fuel; and we have been preoccupied with the running cost of the electricity service process – the number of units of electricity flowing, and the price per unit. We have presumed that we can construct some form of financial equivalence, to link the investment cost of the physical assets involved to the unit cost of the electricity flowing through the system. Even in the context of traditional electricity this approach has been more than somewhat arbitrary, though generally accepted. However, we can now often generate electricity economically close to where it is used. Generator and loads may even belong to the same owner. For such innovative decentralized electricity, focusing so obsessively on the rate of flow of electricity and the price per unit may become irrelevant and even nonsensical, both financially and operationally.

Consider, for example, your iPod. You do not, after all, measure the electricity flowing through the iPod, or pay for it by the unit. You purchase the entire system, including the battery; then you use it as you wish. You may keep track of how long the battery lasts and what a replacement may cost; but you don't have a meter on the iPod measuring the flow, nor do you think of how much the electricity costs per unit; indeed if you did you might be alarmed. The transactions involved when you purchase the iPod and even when you purchase a replacement battery are investments, not commodity transactions. We may eventually come to apply similar financial considerations to larger local electricity systems, at least in certain significant circumstances. We may invest in the requisite electrical assets, and then use them as we wish, with no meter involved.

Infrastructure electricity

As small-scale decentralized generation becomes feasible and attractive, a crucial distinction comes into important prominence. Some generating technologies convert fuel energy into electricity. The process entails using and therefore actually consuming fuel. We can usefully and accurately characterize such fuel-based electricity according to the rate of fuel consumption, and the cost of the fuel per unit. In this case, we pay attention to the flow of electricity and the unit cost of the electricity because it is a direct and appropriate corollary of using the fuel. However, not all electricity generation requires fuel. Indeed, one of the oldest generating technologies was and is hydroelectricity, produced by converting the mechanical energy of falling water into electricity.

For large-scale hydroelectric generation, the quantity and level of water behind a dam represents an analogue of fuel energy, in that the potential energy of the water can be stored and released as desired. In this case, the traditional approach can be justified: you can measure the flow of electricity, equate it to the flow or 'consumption' of water from storage and attribute a unit cost accordingly, just as you do for fuel-based electricity. By contrast, however, for smaller hydro installations, such as run-of-river plants on non-seasonal waterways, the mechanical energy of flowing water is continuous. A mini- or microhydro generator with water flowing through it will produce electricity continuously. If you wish, you can infer the unit cost of this electricity by some form of accounting treatment of the investment cost of the generator. However, you don't pay for the flow of water itself; indeed, provided it is enough to turn the generator you may not even measure it. You cannot usefully identify any 'consumption'.

In this case a physical asset generates electricity by converting natural ambient energy that is already present and costs you nothing. Other innovative technologies such as wind power and photovoltaics do likewise. Generating this electricity involves no commodity transaction at all. When you use this electricity, accordingly, you need not think of a commodity transaction either. For all these technologies, what matters is the investment cost of the physical asset, and how you treat that asset in financial accounting terms. We can call this 'infrastructure electricity'. The implications, especially for small, self-contained electricity systems, especially those based on natural ambient or 'renewable' energy, may be striking. Precedents already exist. If you have the right equipment you can already operate your iPod or laptop computer on rechargeable batteries, and recharge them with solar energy from your roof, with no fuel or commodity involved.

The distinction between fuel-based and infrastructure electricity has prompted thought-provoking analysis to reassess the basis for estimating the cost of electricity. Conventional comparisons now routinely assert, for instance, that electricity from a gas-fired combined-cycle station costs perhaps '2.8 cents per kWh', whereas electricity from a wind farm costs perhaps '4.9 cents per kWh', and that from a photovoltaic array '11.5 cents per kWh', or some such comparative numbers. The policy inference such cost comparisons suggest is that you should prefer the gas-fired combined-cycle station to the wind farm, and that the photovoltaic array cannot compete with the other options.

However, closer examination of the financial techniques used to derive these cost figures reveals serious shortcomings. The groundbreaking work of Shimon Awerbuch, a US financial analyst most recently based at the University of Sussex in the UK, casts serious doubt on the purported costs traditionally used to compare different generating technologies, and the policy conclusions based on these comparisons; see Chapter 6 for more details. In particular, for fuel-based generation, the inferred cost of a unit of electricity depends to a significant extent on the cost of the fuel. Over the operating life of a gas-fired combined-cycle station, for example, twenty years or more, the cost of gas could rise substantially, but unpredictably. To an analyst considering the station as an investment, this fuel-price risk, raising the estimated lifetime cost of electricity from the station and thereby reducing its profits, must incur a penalty, raising the cost of capital. The result may be to double the estimated cost of electricity from the station. By contrast, the cost of infrastructure electricity, perhaps from a wind farm or a photovoltaic array, is known from the outset, an initial capital investment with no fuel-price risk to be considered.

The details of this analysis are still evolving, but some conclusions are already startling. Recent commentary takes for granted that official government support for renewable generating technologies, notably in Europe, increases the cost of electricity to users. However, consider the aggregate generation on a system as a portfolio of investments. If you add low-risk infrastructure generation such as wind farms and even photovoltaics to the portfolio, you actually decrease the overall cost of the electricity generated on the system for a given level of risk, or reduce the risk for a given cost. Who receives the benefit of this reduction in cost is another matter; some participants are clearly accruing returns or 'economic rents' not yet adequately identified. Again, historical evidence indicates that an increase in fossil-fuel prices is correlated to a decrease in economic activity. Electricity prices that go up with fossil-fuel prices will aggravate the economic problem. However, the cost of infrastructure generation is made up almost entirely of the repayment of capital with interest. Reduced economic activity will reduce interest rates, and make the infrastructure electricity less expensive. In this way infrastructure generation adds stabilizing negative feedback and robustness to an economy. This is an attribute we have not hitherto sufficiently acknowledged or analysed. The implications of these innovative financial insights should become a key aspect of electricity policy.

An obvious corollary of the emerging analyses of risk is that so-called 'security of supply' issues for infrastructure electricity look very different from those for fuel-based electricity. The debate about 'energy security' still focuses mainly on the risk of fossil-fuel supply disruption and price volatility. Infrastructure generation eliminates both problems. Moreover, since such generation may be sited closer to users, infrastructure electricity also reduces other risks of traditional electricity with its remotely sited large-scale generation, especially vulnerability to accidental or intentional disruption of networks.

Better business

The financial treatment of fuel-based versus infrastructure generation is only one of many financial issues arising in the transition from traditional to innovative electricity. In traditional electricity finance, revenues come ultimately from charges decreed centrally and paid by captive customers and taxpayers. For networks, at the moment, this approach still broadly prevails, if indirectly. For generation, however, such traditional arrangements were substantially swept away in the first phase of electricity liberalization. Instead generators earn revenues based ultimately on commodity transactions in units of electricity, at unit prices established by a 'market'. Wholesale transactions, mediated by a vast existing infrastructure, particularly of networks, appear to offer a plausible form of business, at least in the short term. However, for all but the most intensive industrial users, competitive retail transactions are yet to prove satisfactory. Vying to sell anonymous units of electricity at the customer's meter, a company can compete only on price. Margins become perilously thin. When customers can change supplier at short notice, a month or less, the customer base becomes hair-raisingly volatile. Even at the wholesale level, problems are now surfacing. Wholesale prices in the UK, for instance, have sometimes fallen so low that some generators cannot even cover their cost of capital. In response, the owners have simply walked away from or shut down power stations, potentially permanently. In such a situation the 'electricity market' could swing from glut to blackout with dismaying speed.

Partly to address this issue, some commentators have suggested the introduction of 'capacity payments', based not on commodity transactions in units of electricity but on a recognition that you must pay for the continuous availability of a physical asset, whether or not it is delivering a service at a given instant. Capacity payments are an early indication of what could gradually become a much more prominent feature of electricity finance, based on investment and contracts for assets, rather than on short-term commodity transactions. This will become especially the case for infrastructure generation such as micro-hydro, wind farms and photovoltaics, as well as for the networks to match.

Asset-based business may also point the way for other transactions and relationships on future electricity systems. Suppliers alarmed by the volatility of customers buying anonymous units of electricity as a commodity will want to win more customer loyalty, and create more long-term relationships. A possible way to do so will be by moving away from commodity transactions towards contracts for services. For instance, your supplier may offer to guarantee you the comfort, illumination, refrigeration and other energy services you desire, on the basis of a contract for a fixed regular payment not determined by a meter, giving you a name and number to call any time in case you have a problem. The supplier is effectively selling you 'peace of mind' – a very attractive proposition, provided the supplier can deliver. The idea is not novel. Some suppliers of on-site cogeneration facilities, for example, have long done business in this way. They sell not the physical plant but its on-site services, on a contract basis. The suppliers install the plant on the site but continue to own it. They do not, however, require any staff actually on the site. Instead they have diagnostic instruments connected to a remote monitoring centre. If you have such an installation, your supplier usually detects any potential malfunction or failure and sends roving technicians to rectify it even before you notice.

Again, a supplier may contract to deliver, say, a guaranteed interior temperature, for a fixed monthly fee. Some UK suppliers, for instance, have launched pilot schemes of this kind, aimed at low-income customers in low-quality housing. This kind of contract for services presents solid economic reasons to get the entire local system right. A supplier with the requisite competence can upgrade building performance, and optimize generation and network together with illumination, motors and drives, fans, pumps, electronics and other end-use equipment, to get the best available return on the combined investments in the entire system – electricity generation, delivery and use.

Such innovative financial and business arrangements could also help to address a fundamental and intractable flaw in energy investment. As noted earlier, if you invest, say, in a power plant whose output you intend to sell, government treats it for tax purposes as a business investment. On the other hand, if you invest, say, in a high-efficiency deep freeze in a household, that is not a business investment. Government accordingly treats it much less generously for tax purposes. This basic assumption, common in taxation regimes around the world, skews investment inevitably towards provision of more electricity generation instead of more efficient end-use equipment. If, however, you integrate the entire local system to sell not units of electricity but energy services, you can legitimately treat the entire system as a business investment. Over time, the consequences for the human energy-service infrastructure of this one single change of policy could be dramatic.

Real energy policy

Selling energy services rather units of fuel or electricity also illustrates a central issue we must address. What we call 'energy policy' is still really 'fuel and power policy', preoccupied with batch transactions in commercial energy carriers that we treat as commodities and price by the unit.

As we noted earlier, the very expression 'energy policy' dates back only to the so-called 'energy crisis' of the early 1970s, when the OPEC oil price rise coincided with problems affecting natural gas, coal and electricity in many countries. 'Energy' became a headline shorthand for all these energy carriers; the term has been used this way in policy ever since. It is profoundly misleading. It implies that one form of so-called 'energy' can be readily substituted for another. In modern industrial society, on the contrary, a particular piece of end-use technology requires a particular specialized energy carrier. A particular car, say, must have high-octane unleaded petrol, not just petrol nor petroleum, much less coal. Changing energy carriers means changing end-use technology too; across society this is an expensive and long-term process. 'Energy policy' that concentrates entirely on the availability and cost per unit of commercial energy carriers misses the most important part of human energy systems – the energy service systems, especially buildings and their fittings.

Real energy policy should be far wider in scope than fuel and power policy. Real energy policy should encompass all aspects of energy investment, not just in facilities to produce and deliver commercial energy carriers, but more importantly in facilities to deliver the energy services people actually want. Real energy policy, therefore, should broaden its vision beyond batch transactions in commodities. The policy levers should include asset accountancy, asset taxation and all the related measures affecting investment in infrastructure of every kind, particularly energy-service infrastructure. Electricity, with its distinctive attributes and its applicability to integrated local systems, could mediate the transition to real energy policy.

At the same time, local electricity systems might also help to counter one infrastructure trend that is now of increasing concern. Mergers and acquisitions are producing a rapid international agglomeration of ownership of electricity and gas networks into a handful of enormously powerful multinational companies. These new monoliths may pose major problems for national regulators, not to mention customers. A monolith may control, for instance, the supply of gas from a remote field. However, you can generate electricity anywhere, at a price, particularly infrastructure electricity. If you dislike the terms on offer from the neighbourhood monolith, you can opt instead for an integrated optimized local infrastructure electricity system, independent of the monolith. At the moment, policymakers usually dismiss such an alternative as too expensive. But more accurate comparative costing, including risks, combined with relentless technical advances, may soon make this alternative seriously attractive. Such complete local systems may indeed be more effective than simple on-site generation as a form of competition to keep monoliths from misbehaving.

If you have sensitive loads such as data processing facilities and server farms, an increasing proportion of electricity use, you will also consider another factor. In the aftermath of liberalization, with reduced redundancy and reduced staffing of fieldwork, electricity networks themselves are now the main source of disturbances such as transients and harmonics, voltage spikes, voltage sags and other potential trouble, not to mention actual power cuts. If you have sensitive loads, such disturbances can be dismayingly expensive. An on-site system using highly reliable generation, especially infrastructure generation, may be an insurance premium you become willing to pay.

Getting there from here

Many if not most of the issues that electricity people now face are complicated primarily by the existence of traditional legacy assets, institutions and mindsets. Some commentators have argued for years that those parts of the world unburdened by the inertia of such a legacy, the two billion people without access to traditional electricity, may be able to 'leapfrog' over traditional arrangements. They may more readily take advantage of the opportunities for innovation in technologies and institutions, and of innovative ways to think about energy in society. Where no stultifying electric legacy exists, people may find that innovative approaches to buildings, to local electricity systems and to infrastructure electricity, untrammeled by problems with fuel, may be both more necessary and easier. Nevertheless, such leapfrogging can only take place with the active involvement of those who already have electricity, the technologies, the experience and the finances. The potential for mutually beneficial cooperation is vast; but the stumbling blocks are many and obtrusive.

Much contemporary policy on sustainability, on energy and especially on electricity, focuses on reducing the environmental impact of providing and using energy, particularly the impact on climate. For at least three decades governments have been exhorting their citizens to use less energy. This is simply erroneous. We do not need to reduce the use of energy; we can use as much energy as we wish. We need to reduce the use of *fuel*, a very different and much narrower problem. We need to use less fuel, and more infrastructure energy – not only the familiar and unnoticed infrastructure energy in the form of heat, but also innovative infrastructure energy in the form of electricity. Indeed that approach may nevertheless reduce overall energy use anyway. Concentrating on infrastructure electricity in local systems will encourage optimized design, with high-performance loads to take full advantage of local electricity. Some commentators speculate as to what 'sustainable electricity' might look like. As infrastructure electricity – unmetered, unmeasured, paid for as an investment rather than by the unit – gradually becomes a larger proportion of the mix, sustainable electricity may eventually be completely invisible.

We must, however, get there from here, around the world, while minimizing disruption and keeping the lights on. The prize could be a sustainable planet; but the task is awesome and daunting. It will demand technical flair, financial acumen and political courage. Can we meet the electric challenge?

6

GENERATING CHANGE

Why generate electricity? The question seems absurd. Electricity keeps the lights on, and makes possible countless other services. Those of us in the fortunate parts of the world take electricity and its uses completely for granted. Refine the question, however, and it takes on surprising urgency, at least for some of those involved. Why generate electricity in this way, in this form? Is this way the best way? Is this form the best form, given what we expect of electricity, and the opportunities now available? A decade or so ago these questions too would have looked absurd. Now the answers are anything but obvious. We are no longer certain which technologies to choose to generate, deliver and use electricity, or where to install them, under what ownership or control. Indeed we no longer know who should decide, or how. We have begun to break away from traditional electricity, and from the centralized monopoly franchise with captive customers; but we have yet to find a wholly satisfactory alternative. Electricity business used to be unchanging, even boring, safe enough for widows and orphans. It has now become almost as nerve-racking as bungee-jumping.

The cumulative impact of so much uncertainty raises an immediate corollary. If we were starting now, with no preconditions, knowing what we now know, to design a system to generate and use electricity, it would almost certainly look more or less different from all the systems we actually have. In particular it would look distinctly different from the central-station synchronized alternating current system, replicated over much of the world in the previous century. Yet this type of system now includes some 3000 gigawatts of generating capacity, and has served billions of electricity users well for half a century. It may no longer be the best we can do, or even good enough; but it has demonstrated unequivocally the remarkable utility and versatility of electricity and the

This chapter is adapted from 'Generating change', Working Paper 2 for *Keeping The Lights On*, Chatham House, 2003.

services it delivers. People all over the world rely on it. For those without it – two billion people, remember, one-third of humanity – life is more difficult, often much more. What we do with electricity, the systems we already have and those we need, will be crucial for sustainable development, for climate, indeed for the future of humanity on Earth. How may existing electricity systems change, and why? What may new systems be like, and why? The way we answer these questions will have sweeping implications for technical configurations, financial arrangements, institutional frameworks and business relationships.

To understand change that may happen, we can start by revisiting change that has happened already. The transition now under way shows some intriguing parallels with the transition that took place in the early decades of electricity. Then, as now, the issues for generation included choice of technology; cost; size; location; ownership; connection to networks; source of revenue; and planning constraints. Many of the criteria then invoked are still in use – no longer always defensibly. Indeed the most dramatic changes in electricity systems happened more than a century ago, between the 1870s and the 1890s.

In the 1870s, an entrepreneur such as Thomas Edison sold complete systems, for electric arc lighting. If you were the customer, you took title to all the assets – generator, cables, controls, arc-lamps – all of which were installed on your site. You were then responsible for operating and maintaining the whole system – that is, for keeping your own lights on. In 1882, however, as we noted earlier, Edison used his new incandescent lamp to establish central-station systems, in Holborn, London, and then in the Wall Street district of lower Manhattan. At the outset Edison retained title to all the assets, and charged his customers according to how many lamps they used. He was selling illumination, the service that the complete electricity system delivered. Then, by the mid-1880s, as noted earlier, the first practical electricity meter arrived, to measure the flow of electricity through a circuit. Within a very short time Edison, his competitors and other electric entrepreneurs in North America, Europe, Japan and farther afield, were selling not arc-lighting systems nor incandescent light but electricity – the process – measured and priced by the unit.

Scaling up

The key reason for this rapid change in the nature of the electricity business, and the relationship between suppliers and customers, was the cost of generating electricity, and the possibilities then available to reduce this cost. Not surprisingly, that has remained a determining factor in the evolution of electricity ever since. In the 1870s, if you wanted to use electricity, you had to choose between the two available practical options for generation, the battery or the rotating machine called the 'dynamo', spinning a wire in a magnetic field to make electric current flow in the wire. Telegraph companies used batteries, because they could be sited where needed and readily replaced, and because the system did not require large flows of electric current. Arc light, however, required a substantial current; to provide it with batteries would be expensive and inconvenient. You could use a dynamo to deliver a much larger current continuously with comparative ease; but you had to turn it with a 'prime mover' – a source of kinetic energy to be converted into electrical energy. In the 1870s you could choose between two kinds of prime mover, the water wheel and the steam engine. A water wheel could use the energy from a flow of water, a stream or waterfall, that was effectively free of charge; but you had to site the wheel, and the dynamo it turned, where the water was – not necessarily close to where you wanted the light. On the other hand, you could site a steam engine anywhere, provided it had access to a supply of fuel, almost invariably coal. With a steam engine and dynamo you could have light essentially anywhere – anywhere, that is, that you could tolerate the noise, smoke and ash from the generator. You did, however, have to buy and pay for the coal to generate the electricity.

These two forms of generation, using water power and steam power, typified a dichotomy that persists into the twenty-first century. Its significance for the future of world electricity will steadily increase. If you use water power to generate 'hydroelectricity', you use an array of physical assets to convert a flow of natural ambient energy, that of falling water, into electricity. The water flow is a natural phenomenon. It is delivered by and dependent on rain and topography. It continues whether or not you use any of the flow to turn a rotating machine to generate electricity. The cost of electricity from this kind of generation depends almost completely on the investment cost of the physical infrastructure that converts water energy to electrical energy. By contrast, if you use steam power to generate electricity, you have to have fuel to raise the steam. The cost of the fuel may represent a substantial fraction of the total cost of the electricity produced. We need to recognize the distinction between such fuel-based generation and infrastructure generation that does not require fuel, including not only hydroelectricity but wind power, wave power, tidal power, solar thermal and photovoltaics. We must scrutinize much more closely the implications of this distinction, not least for the finances, the risks, and the environmental impact of electricity.

In the early decades of electricity, both water-powered and steam-powered generation, while differing markedly in other respects, shared one crucial attribute: economy of unit scale. Water-wheels and water turbines, steam engines and steam turbines, dynamos and alternators all exhibited this desirable feature: a larger unit produced cheaper electricity. This was why Edison and his contemporaries pursued the concept of central-station generation, distributing electricity over an extended network. Only by increasing the size of the generator could they make the cost of electric light competitive with that of gas light from town gas. Such a large generator would produce more electricity than could be used on a single site; it needed a network through public space to deliver the electricity to users on other sites. However, the cost savings on a larger generator would more than make up for the extra cost of the extended network. The entire concept and configuration of central-station electric light was based directly on the already successful model of central-station gas light, and for much the same reasons, primarily economy of scale of production.

Generating loads

One major difference arose, however. Town gas was a physical commodity, produced by roasting coal in large retorts. You could store town gas, in huge expanding tanks called, for some reason, 'gasometers', that served as a buffer between the retorts and the users of the gas. Although people wanted illumination mainly in the hours of darkness, the retorts could operate continuously at maximum output, filling the tanks during the daytime to help meet the load at night. Electricity, however, was not a commodity but a physical process, happening simultaneously and instantaneously throughout the entire interconnected system. You could store gas, but not electricity. You had to generate it essentially at the instant your customers used it. The system therefore had to have enough generation available to meet the maximum load on the system. For most of the day, however, this generation would operate below its maximum output or even stand idle.

For a self-contained arc-lighting system, this was not really an issue. If you were the owner and operator, you simply switched on the generator when you wanted light, exactly as you still do with a hand-held electric torch or flashlight. As owner, you accepted the cost of having the generator and the rest of the system idle for much of the time, as a corollary of having it available when you wanted light. You did not measure the flow of electricity, much less pay for it by the unit. If, however, you were the operator of a central-station system, you had to install and operate generation able to meet the maximum potential load; and you had to recoup the investment from customers. But you had to keep charges down; customers unhappy with the cost of electric light might revert to gas light. On the other hand, generation operating below capacity was earning less revenue; if it was standing idle it was earning none at all, though you still had to pay capital charges.

The mismatch between maximum load and maximum generation, such a headache for early electric entrepreneurs, is an inevitable corollary of breaking the continuity of the electricity process – of treating electricity as a quasi-commodity and selling it by the measured unit. The customer purchases electricity by turning on a switch, and expects the electricity to be delivered essentially instantaneously. The system must respond accordingly, increasing generation to match the increased load. According to this arrangement, the loads are ready and waiting to be connected to the system whenever the independent owners of the loads so desire. The system in turn must have generation ready and waiting to respond. This state of affairs, the fundamental tenet of the traditional central-station electricity system, has prevailed for so long that it feels sacrosanct. It need not be.

Edison, his contemporaries and their successors tackled the problem in three ways. They expanded systems, they added new types of load, and they interconnected systems. If you expanded the system by adding more customers and more loads, you could justify installing ever-larger generators; the consequent economies of scale steadily lowered the cost of generating electricity. If you could add loads on premises already connected, avoiding the need to expand networks, so much the better. Introducing a different type of load, the electric motor, created uses for electricity during the daylight hours when few lights were on.

Other substantial changes quickly ensued. Edison and fellow entrepreneurs assumed at the outset that the central-station systems being set up would generate and deliver so-called 'direct current' or DC, always flowing in one direction like that from a battery. By 1890, however, George Westinghouse and Nikola Tesla had shown that alternating current, AC, surging rapidly back and forth in the wires, had several marked advantages over DC. You could generate AC easily with a rotating machine. AC could power an elegant motor. You could deliver AC with a more economical arrangement of conducting cables. Perhaps most important of all, AC allowed you to use 'transformers', to increase or decrease the voltage, and thereby to decrease or increase the current in proportion, easily and efficiently.

This offered a solution to a major problem of DC. Electric current in a wire heats the wire, wasting energy; and doubling the current quadruples the heating effect. The longer the wire the worse the losses. Central-station systems like those of Edison, based on DC, were severely hampered by energy loss from the network; the more widespread the system the worse the losses, making system expansion prohibitively costly. To reduce the energy loss you must either use heavier, more expensive wire or a smaller current. For a given flow of electrical energy in a given wire, doubling the voltage halves the current. By making the voltage higher, you can make the current lower, and therefore reduce the resulting loss of energy. Accordingly, by introducing a suitable transformer into an AC circuit you can raise the voltage high enough to reduce the current and the corresponding losses dramatically. This allows you to make your line much longer while still being economic to operate. At the other end of the circuit you can use another transformer to bring the voltage back down to a level suitable for whatever loads may be connected.

Expanding monopolies

The advent of AC dramatically multiplied technical opportunities. Then, gradually, as entrepreneurs added more generators and extended delivery networks, interconnecting systems became feasible and attractive. For generation, interconnection meant reducing the maximum capacity required to allow for faults or failures at times of maximum load. Moreover, if systems had differing proportions of motors or lighting, differing 'load profiles', linking the systems improved the overall utilization of total generation. From the 1890s onwards the so-called 'load factor', the actual output of system generation as a fraction of its potential maximum output, became a major concern. Since revenue depended on the number of units of electricity sold, load factor became a key indicator of return on system assets.

In the early years, when systems were still small, skewed load profiles led to low load factors and seriously hampered economic performance. Electric entrepreneurs embarked on a process that continued for decades, called 'load-building', intended not only to increase the total use of electricity, and the generation needed to provide it, but more particularly to increase the use of electricity in daylight hours, to balance the lighting load in the evening and night-time. Load-building had another corollary, less obvious. The use, and the uses, of electricity expanded into more and more aspects of daily life. Applications once confined mainly to industry, especially motors, became ever more common in households. Electricity use on weekends steadily increased. Society began to regard electricity – not just electricity, but specifically electricity generated in central stations and delivered over a network – not as a luxury but as an essential.

That set the stage for a crucial step in the evolution of electricity systems: establishing the monopoly franchise. Until the 1920s, central-station generators had to compete not only with gas for lighting but also with on-site and other local electricity generation, notably by tram companies and factories. But central-station proponents such as Samuel Insull of Commonwealth Edison in Chicago, along with many others, lobbied politicians relentlessly, claiming that an electricity system was a 'natural monopoly'. At length the relevant politicians in country after country agreed. Within the monopoly franchise area only the franchise holder could generate electricity to sell. You could generate and use your own, but you could not sell any surplus. If you wanted to buy electricity you had to take it from the monopoly supplier in your area. The regulatory authority appointed by the relevant government decreed the terms.

The monopoly franchise meant that the captive customers of the system bore all the risks, including those of investment in system assets. This in turn allowed suppliers to build ever-larger generating units. In principle, such larger generators were expected to produce cheaper electricity. But even if they did not, no matter what they cost or how long they took to commission, the captive customers had to pay. This gave the central-station system an advantage that slowly, inexorably suffocated on-site generation. In urban areas, local generators shut down their own plants and began to buy their electricity from the monopoly system. In rural areas such as the midwestern states and provinces of North America and outlying areas of Scotland, small-scale privately owned on-site wind generation briefly burgeoned. But monopoly systems, massively subsidized by taxpayers, extended their lightly loaded networks into these areas and snuffed out the local generation. Natural or not, the monopoly franchise became an established political reality, to be taken for granted for more than half a century. From the 1920s onwards, the monopoly franchise, under some form of government regulation, became the basis of electricity systems around the world. It remained so until the late 1980s and the advent of liberalization. Even after liberalization, networks have usually remained monopolies, not always for good reason.

Expanding generators

The monopoly franchise fostered a steady increase in the unit size of individual generators and of the central stations housing them. From the 1950s through the 1980s the scale-up was relentless. As steam-cycle materials and engineering steadily improved, fuel-based generation gradually overtook and then outstripped hydro generation. Coal remained the dominant fuel for raising steam. However, as oil refineries sprang up to service the expanding population of internal combustion engines, heavy residual oil became both abundant and cheap. Many systems added large generating stations burning heavy residual oil, as adjuncts to refineries. Natural gas, a troublesome accompaniment emerging from some oil wells, also became available as a fuel to raise steam, provided you had a nearby network connection to justify investing in a generator rather than simply flaring the gas to get rid of it, and provided the monopoly system would accept your electricity. From the mid-1950s onwards, national governments in many countries vigorously promoted and financed a yet more exotic way to raise steam, using nuclear reactors in nuclear power stations.

Units became so large that system operators had to keep major turboalternators turning at the network frequency, 'synchronized' but delivering no power, as so-called 'spinning reserve', ready to take over in seconds if an operating unit should fail. Maintaining the stability of vast and intricate alternating-current networks required a whole catalogue of such so-called 'ancillary services'. On traditional monopoly systems, generators delivered these ancillary services as a matter of course, as part of the centralized operating regime. The additional costs to generators of providing the ancillary services went largely unrecognized.

By the 1960s the range of options for generation, especially fuel-based generation, presented system planners with difficult choices. But one overriding consideration prevailed almost everywhere, right into the 1980s: no matter what mishaps, misjudgements and mistakes they suffered, the planners did not have to answer for them. The monopoly franchise saw to that. Moreover, during this time power stations had grown so large that they routinely took a decade or more to plan, build and bring into operation, by which time those initially responsible had not infrequently retired or even died. Planners nevertheless continued to cite the anticipated cost of the electricity output from a new station as the guiding criterion for their choice of generating technology and fuel. For new nuclear stations, pronouncements about anticipated cost of electricity output plumbed depths of absurdity not hitherto observed.

Comparing generation

From that time on the question of cost, as applied to electricity generation, should have attracted much more critical study than it has subsequently received. Even since liberalization, electricity planners continue to cite the purported comparative cost of different forms of generation as the basis for choosing generating technologies and fuels. In a liberalized context, however, those making the decision, those choosing generation in which to invest, are no longer immune from the consequences of their mistakes. They are betting their own money, and that of their shareholders and bankers. Their track record since liberalization is unimpressive, to put it mildly. Yet the catalogue of options for generation is now longer and more varied than ever before. In the new liberal context, the problem is further complicated by uncertainties about network costs including connection, about electricity wholesale prices, and about level of demand. Moreover, generating units on a single interconnected system may have a variety of different owners, with widely differing business agendas and reasons for owning generation. Valid selection criteria for new generating capacity, including persuasive and accurate comparative costings, have never been more necessary, nor more elusive.

Start with the ideal. What might ideal electricity generation look like? It would be imperceptible. The light would go on, the motor start, the computer function, with no visible corollary anywhere, no cost and no side-effects. Electricity itself is indeed imperceptible; but existing generation, especially traditional generation, is anything but. Generating technology costs money, often a very great deal of money, in investment and – especially for fuel-based generation – running expenses. Generators may be visually obstrusive, sometimes dramatically so. They may be noisy, smoky and smelly, and discharge quantities of waste, gaseous, liquid and solid. They may also emit greenhouse gases, particularly carbon dioxide. At the end of their useful lives they may prove difficult to remove. Ideal electricity generation would have to eliminate all these drawbacks.

Perhaps the closest actual approximation to ideal generation might be a small hydroelectric station built, say, in the 1930s. Its capital cost is long since amortized, and it incurs no fuel cost. It requires no permanent staff and minimal maintenance. Its impact on the environment has largely been accommodated and assimilated by its surroundings. It is part of the built infrastructure; but over the decades it has become in effect part of the landscape, a part of the landscape that generates electricity. The only visible corollary of its function would be the network carrying its electricity output to loads somewhere else. If the cables were buried, the presence of the generation, and indeed of the entire electricity system for this generation, would be even more subtle. Lingering questions about the reliability, power quality and vulnerability of electricity from this generation are likely to arise more from the delivery network than from the generation itself. If the network can be shortened and simplified, these residual questions dwindle almost to insignificance.

The only substantial question, admittedly non-trivial in some cases, relates to the flow of natural energy, in this case the water flow, that drives the generator. If the flow subsides or ceases, so does the output of electricity. That is an inevitable corollary of such infrastructure generation, and an argument for interconnecting infrastructure generation of different types and locations, to minimize the circumstances in which natural energy flows may prove insufficient to provide the electricity desired.

Unfortunately, adding a new hydro station built sixty years ago is not an option. Ideal generation must remain an ideal, not a practical reality. But comparison of the ideal with the practical reality is nevertheless instructive. Could we gradually approximate to the ideal, over time, from where we are now? How might we do so? Having an ideal in mind as a long-term objective could provide a useful touchstone, and even indicate potential strategic pathways, particularly when the catalogue of potential generating options is already long and growing longer. We must, however, acknowledge key corollaries of any such hypothetical analysis. Before liberalization, electricity planners happily laid out schemes extending at least four decades into the future; some of the wilder nuclear schemes were even longer. In liberalized electricity, however, no such central planning function any longer exists. Nothing even in principle can implement any long-term programme of convergence towards an ideal. Moreover, individual system participants have widely differing interests and agendas. They tend to be at best medium term, far shorter than the timescale of possible convergence towards any ideal, which will be decades, conceivably most of this new century.

Independent loads, dependent generation?

These considerations have a further profound implication for future generation. Traditional electricity accepted as a premise that the total load, and the individual loads, on the system were independent and autonomous. The system had to provide generation to match this independent load. Under a monopoly franchise with captive customers the premise was tenable, especially when the requisite generation came in large discrete lumps readily amenable to central planning. In a liberalized context and with today's technology options, however, the premise that generation has to expand to match independent load is a misconception. Indeed it is potentially a dangerous delusion. Governments and their agencies continue to present scenarios and analyses purporting to demonstrate the 'need' for new generation, to meet postulated increases of load. However, short of renationalizing electricity systems, the scenarists and analysts have no direct way to implement proposed plans for, say, new nuclear power stations, or any other form of generation. The owners and operators of other generators on a liberalized system would be outraged if government itself undertook to construct new generation that would compete with existing generation privately owned. Indeed private generators are already up in arms about indirect government support and preferential treatment, including tax breaks, grants and other financial assistance, given to one generator or one form of generation and not to others. Controversy rages, about comparative subsidies to fossil-fired, nuclear and renewable generation in various places.

We need, however, to ask a more fundamental question. Why should generation per se qualify for government support that is not made available to loads? Why does electricity policy continue to assume that loads are autonomous, but that generation is not? Historically, generators were much larger and much less numerous than loads, and either directly or closely overseen by government. For straightforward practical reasons, generation was therefore much more readily amenable to government policy intervention. As systems evolve, and generation becomes smaller, more numerous, more variously owned and more decentralized, in fact increasingly similar to loads, the asymmetry of policy focused primarily on generation is going to become egregious.

Many governments have recently expressed strong opinions about the future of electricity, seeking policy measures to achieve 'security of supply', affordable electricity services, and environmental and social sustainability. Henceforth, however, governments must recognize that electricity policy will have to rely on less direct methods to influence the evolution of electricity systems. Instead of planning and implementing system expansion and change explicitly and centrally, governments will have to establish a fundamental framework within which front-line electricity decision-makers – generators, network planners and operators, and electricity users – must function. Such a framework, already well within the accepted powers of democratically elected governments, will have to evolve significantly from that which is now tacitly in effect in even the most liberalized electricity regimes. Moreover, it will have to embrace not just generation and networks but complete systems, including loads.

Governments may decide, for instance, to impose minimum technical and physical standards, and indeed a straightforward ban on electric technologies, including end-use technologies, with inadequate performance. They may widen so-called 'energy taxation' beyond fuels and electricity to embrace also, and more particularly, taxes on energy assets – not only power stations and networks but also buildings and appliances. Governments may also have recourse to many other financial measures, such as grants and early depreciation, as well as procurement and regulation. They should revisit and revise existing subsidy regimes for electricity assets and electricity fuels, to foster developments they desire, and discourage the undesirable. Over time, they should widen such policies and policy instruments to embrace symmetrically the entire interconnected system, including in a balanced way both the generators and the loads that may or may not be connected to the system at any given instant.

One obvious priority will be to shift away from policies preoccupied with short-term commodity transactions in flows of electricity priced by the unit. Instead, policies will have to encompass whole circuits and systems, not only generation but also networks and loads, and treat whole systems as coherent arrays of assets that function together to deliver services. In an appropriate business transaction you will then, for instance, pay a fixed fee to be able to use electricity assets as and when you want to, without referring to measured or metered flows of electricity through them. As noted earlier, a particular example of this approach is the so-called 'capacity payment' to keep generating capacity available when the system does not immediately need it. In suitable circumstances governments and regulators can generalize this approach, so that electricity customers pay fixed fees for services they then use as and when they wish.

At the same time, some commentators also expect the emergence of local generation and local systems to engage users themselves more actively in providing and using electricity services. Historically, electricity and its services have been delivered by someone else; you the user are a passive recipient. Most government policy still makes this assumption. However, when your own generation is keeping your own lights on, you may pay more attention to the electricity you use, why you use it, how much and when. That in turn may prompt you to put more effort, and indeed investment, into your own facilities, to ensure you get the best performance out of your premises, fittings and appliances. Such personal involvement will be more effective than any amount of official policy, however enlightened.

Transitional electricity

Needless to say, where traditional and what we might call 'post-traditional' or 'transitional' electricity systems already exist, any radical change is going to be both piecemeal and gradual. Around the world, electrical loads in thousands of gigawatts expect to have electricity available as before, without interruption or dislocation. In many instances significant interruption could threaten lives. The very continuity of electricity use makes any change in system characteristics, especially technical characteristics, demanding and difficult, even when the whole system except the loads belongs to a single owner and operates as a monopoly. For a post-traditional, transitional system, maintaining the system in continuous stable operation while changing technical characteristics multiplies the difficulty.

For those in charge, the simplest answer is not to change; and many owners of traditional generation are likewise determined to keep things more or less as they have been hitherto. Hence the running battles now raging in many places about connection and operation protocols for small-scale decentralized generation and cogeneration, local and on site. Until recently, such local and on-site generation has relied largely on diesel engines, raising issues of pollution and noise as additional obstacles. Gas engines are more acceptable; so are microturbines, Stirling engines and fuel cells, all of which are now or soon will be commercially attractive. But such local generation still faces stubborn opposition in some quarters. Although couched in technical terms, the battles are actually about comparative financial advantage in a competitive context. That does not make resolving them any easier. The possibility, indeed, exists that the most rapid and sweeping change in how we think about and organize electricity may happen where no 'legacy' assets or institutions impede the change, in the parts of the world where two billion people are still waiting for electricity and the services it makes possible. Unfortunately, however, all too often the authorities in these countries have an even more traditional view of electricity than that of their Organisation for Economic Co-operation and Development (OECD) counterparts, and are a tempting market for outdated technology. For this and other reasons, the political issues arising from changing electricity arrangements essentially anywhere in the world dwarf the technical issues.

The politics of changing electricity start with a straightforward key question: who now decides? Who decides what new system facilities to build, when and where, and why? Who decides which facilities to operate, and when? In traditional electricity, the owner-operators of generating plant and networks decided how to operate and how to expand the system, under the direct or indirect oversight of the relevant government or government-appointed regulator. The planners forecast the future growth of electricity use on the system, and added more facilities, generating plants and networks accordingly, to meet the increased use anticipated. In selecting technologies, they used techniques of comparative investment analysis that routinely failed to anticipate costs accurately, or even approximately. Politically, this often caused some embarrassment; but central planners were much more concerned to avoid the political fallout of blackouts or power cuts. Surplus and redundant generation and network capacity was insurance against such political fallout; and captive customers paid for it.

In the post-traditional transitional context, however, both the decision-makers and the criteria they apply have changed fundamentally. The decision-makers are likely to be private companies, often multinational, with wide-ranging holdings of assets, often in several different countries and on several different electricity systems, not necessarily interconnected. Neither the management nor the staff have any particular dedication to keeping the lights on. Their overriding interest is the entirely respectable objective of returning a profit to their owners and shareholders. The criteria they apply to investing in and operating electricity assets are those they think appropriate to fulfil this objective. In the case of generation, in the first decade and a half of liberalized electricity, the choices they have made have been at best only intermittently successful.

At liberalization they inherited many existing plants based on traditional technologies and fuels, especially coal-fired and nuclear steam-cycle plants. For investing in new plants, however, they have largely abandoned this tradition, opting instead for gas turbines burning natural gas, usually in combined-cycle stations. If liberalization did not precisely bring about this change of generating technology, it certainly encouraged it. Indeed owners shut down many of the inherited traditional plants, sometimes modern, fully functional and not yet amortized.

In the early years of liberalization, privatized operators with extensive portfolios of inherited generation, notably in the UK, operated their units, not just plants but individual turboalternators, in ways that maximized the overall return to the company. Instead of running all available generators all the time, they withheld some from the 'pool' of generation. The effect was to reproduce within the generating company a subset of the traditional 'merit order', when all generation on the system belonged to the same owner and individual generators were called on in order of operating cost, the cheapest first. Withholding generators from the 'pool' reduced the margin of surplus capacity on the system, and therefore raised the wholesale price of electricity. As a result the generators actually operating earned more revenue for the company per unit of electricity output.

At the time the procedure was entirely legal; but the liberalizers, making up the market rules as they went along, nevertheless decried the companies for 'gaming the system'. Partly because of consequent changes in regulation and market rules, and partly because the original privatized companies have themselves undergone so many changes, owners of generation, even large multinationals, now tend to concentrate on individual plants as profit centres. The usual system constraints, however, still apply. Any given unit can operate only according to the overall requirements of the system, however determined, dispatched either by traditional merit order or on the basis of some market. Many generating units operate, of necessity, only part of the time, sometimes such a brief part that the unit does not even earn its cost of capital.

This was always true for a traditional system, with costs and returns aggregated and averaged across the portfolio. On a liberalized system, however, it is already causing trouble. Each individual generator is paid according to how many units of electricity it delivers into the system. The owner therefore wants it to operate at full output all the time, to earn as much revenue as possible. Generators already in existence and in operation on a system do of course follow load and respond to peak loads, mainly because for many such generators this is the only way the system allows them to operate at all. No one, however, wants to invest in a new generator to follow load, much less to operate only at times of system peak load; market prices for electricity are too unpredictable and the investment therefore too risky.

A traditional system has a portfolio of generation in which each unit has a role to play in meeting the overall load at any given time, with technology and fuel appropriate to that role. Since liberalization, however, the traditional portfolio has given way to a default situation in which the essential function of load-following at any given moment devolves onto whichever plants and units happen to be left after all the rest are operating continuously – whether or not these leftover units are technically suitable for the function. So long as the system continues to include a 'tail' of such plants and units, whose owners are willing to keep them available for intermittent operation, system stability can probably be maintained. But owners are already withdrawing from markets, writing down the value of generating assets they consider underused, mothballing or even retiring them. Under such conditions a system can go from an apparent glut of generating capacity to a shortage with dramatic speed.

Electricity users in general want reliable services and lower bills. However, the aim of the liberalizers and the competitive market in units of electricity, acclaimed triumphantly over and over, is ever-lower prices per unit of electricity. This is not the same as lower bills; indeed recent research by Aviel Verbruggen at the University of Antwerp suggests that a higher unit price, encouraging more efficient use, may actually lead to a lower bill for the same service.

In any case, sooner or later, unless the obsessive aim for a lower unit electricity price is modified, it may become incompatible with maintaining adequate reserve generating capacity even to meet peak load, to say nothing of fault conditions. One way to modify the market, for instance, would be to pay owners directly to keep generators available for use, even when not used – the capacity payment mentioned earlier. Market purists decry any such measure. They claim it dilutes the market forces that are supposed to drive the participants and determine their policies and their decisions. But a pure market in ephemeral units of electricity may not indefinitely be capable of keeping the lights on.

Costing generation

How generators should be paid is not the only financial question mark hanging over the future of generation on liberalized electricity systems. The costs of generation are at last, belatedly, coming under the kind of scrutiny they should have received long since, both before and after liberalization. As noted earlier, the analyses of costs, particularly investment costs, for traditional generation on traditional monopoly systems was demonstrably unsatisfactory for decades, to put the matter no more strongly. Since captive customers with no recourse paid the costs in question, cost analysis remained more a matter of wishful thinking and public relations than a serious discipline. Planners decided what generation they wanted to build, when and where, and produced cost estimates to support the decision; but the cost estimates often looked both post hoc and arbitrary.

Liberalization exposed the shallowness of such estimating techiques. Even for existing generation, the purchasers of power stations being privatized found themselves not so much in a market as in a lottery, in which they might earn a substantial return on the investment, or might instead be forced to write off hundreds of millions of dollars or pounds on a plant unable to pay its way at all. Investment in new generation proved to be yet riskier. In the UK, for instance, after liberalization, entrepreneurs invested in an assortment of new combined-cycle gas-turbine (CCGT) stations, burning natural gas purchased on long-term contracts at what then appeared to be a bargain price. These stations forthwith undercut most of the existing traditional steam-cycle stations on the system, including several very large modern stations, forcing them out of business. Within five years, however, a subsequent wave of CCGT stations came into operation, using yet newer gas turbines, more efficient and cheaper, burning natural gas at an even lower price. The first wave of CCGT stations, still less than five years old and far from fully amortized, in turn found themselves being crowded off the system by generation cheaper still. For those whose objective was a lower unit price of electricity this was evidence of the success of liberalization. For many electricity entrepreneurs and investors, losing their shirts, it was evidence that electricity was now a business they did not want to be in.

In an interconnected electricity system, not only the revenues but also the costs of a particular generator depend to a significant extent on the rest of the system and how it operates. To give but one obvious example: if the system load and other generation make a given steam-cycle unit operate at below maximum capacity, as is often the case, the unit's fuel efficiency falls, and its output therefore costs more per unit. Against this background of continually shifting non-linearity, the common practice of stating the 'cost' of a unit of electricity as '2.7 cents per kWh' or some similar figure is frankly indefensible. It becomes yet more so when such numbers, stated even to three significant figures, are used to advocate or justify choosing to invest in a particular generator technology or design as against others claimed to produce 'more expensive' electricity. The practice was disreputable even when the choice lay between otherwise similar technologies, as for example between types of coal-fired or nuclear generation. When the choice is between technologies so fundamentally different as, say, gas-fired combined cycles and photovoltaics, the use of such purported cost comparisons becomes egregiously tendentious.

In any case, moreover, recent studies suggest that traditional techniques to estimate the anticipated cost of electricity from a proposed generator may be inherently and seriously flawed. In 2002, as summarized in Chapter 4, Shimon Awerbuch, an American financial analyst then acting as senior advisor in the Renewable Energy Unit of the International Energy Agency in Paris, produced a report called *Estimating Electricity Costs and Prices: The Effects of Market Risk and Taxes.* The report demonstrated just how untrustworthy such estimates can be. Awerbuch's analysis, which he then dramatically expanded in many parts of the world, is straightforward, if complex to demonstrate. It declares that the traditional approach to estimating the cost of electricity from a particular generator is based on engineering economics rather than financial economics. Engineering economics fails to apply a premium to account for the risk that, over the life of the generator, fuel prices and fuel taxes may vary from those used to estimate the cost of electricity. So long as alternative generating options have broadly similar risks, and those risks move in the same direction with contingencies, the effect on choice of generating technology may be modest to trivial. However, between technologies with dramatically different risk profiles, failure to account for risk may drastically skew the comparison of costs.

Consider, for instance, comparing fuel-based generation with non-biomass renewable generation – say, a gas-fired combined-cycle station with a wind farm. An investor trying to choose between putting money into one or the other will be aware that the price of natural gas may rise unpredictably during the operating life of the combined-cycle station. The investor will therefore require a higher return, to compensate for the risk that the station output may not be as profitable as anticipated. That in turn will increase the cost of generating a unit of electricity. For the wind farm, however, no such fuel-price risk arises. Apart from small and predictable running costs for maintenance, the entire cost of the wind farm is the initial capital investment, known at the outset and unvarying over the operating life of the wind farm. Using well-established techniques of financial analysis demonstrates that adding renewable generation free of fuel-price risk to a generating portfolio otherwise based on fossil fuels reduces the risk for an equivalent return, or alternatively increases the return for the same risk.

Again, an increase in fossil-fuel prices appears to be strongly correlated with a downturn in overall economic activity, reducing demand for electricity and aggravating the problem of higher electricity cost. Renewables, however, whose costs are mainly repaying capital with interest, may actually benefit from the economic downturn, if interest rates fall. Adding renewables thus diversifies the portfolio and reinforces its robustness against unwelcome surprises. The prevailing assumption is that official support for renewables, especially in Europe, is making electricity more expensive. The financial reality, however, may well be that adding renewables free of fuel-price risk should reduce the overall investment cost of generation on systems. Developing and extending this ground-breaking comparative analysis of generating options, refining and sharpening estimates of comparative cost in this way could have striking consequences for the technology choices that drive the evolution of electricity systems.

Comparing environmental impacts

Other aspects of comparative generating cost are likewise controversial. For instance, environmental impacts associated with different forms of generation have been called 'externalities' because their putative costs are borne not by the generator but by the environment within which it operates – the air, the land, the water, and by extension the other people who use the same environment. The decision as to whether and how to account for such externalities has a dramatic effect on the cost, operability and profitability of individual generating plant. Over the years analysts, planners, legislators and regulators have tried to quantify these externalities, and incorporate some suitable numerical and financial measure into the costs attributed to generators. The judgments are necessarily arbitrary; some consider them invidious. Comparative quantification, perhaps in cents or pence per unit of electricity, of the different environmental impacts of, say, coal-fired, nuclear or wind-powered generation is ultimately political, not scientific.

The most successful approach thus far has been so-called 'cap and trade', as applied to sulphur emissions from fossil-fired plants in the US. But the crucial step is the political one of imposing an upper limit on total permitted emissions, and allocating permits accordingly. Only thereafter does the market in permits come into play. The political arm-twisting involved is yet more apparent in the so-called 'grandfathering' of the oldest and dirtiest coal-fired plants in the US. Allowing them to operate as before makes them accordingly much the cheapest generators to run, because they do not have to pay for their externalities. This in turn keeps cleaner and more environmentally acceptable plants off the system.

The US market in sulphur permits has nevertheless served as a model for a similar market in permits to emit carbon dioxide – not in the US but in Europe. The UK pioneered a so-called 'emission trading system' or ETS; the EU then launched a much more wide-ranging ETS. But the 'national allocation plans', according to which national EU governments define the caps to be imposed on their national emissions, have thus far proved so generous that the companies nominally constrained have in practice been compelled to do almost nothing to comply. As a result the price of a carbon allowance remains too low to elicit much reduction of emissions. The second phase of the EU ETS must impose genuine limits, and enforce them, if emissions trading is to have any impact whatever on climate change. Many people, including companies specifically making a business of carbon trading, are devoting time and effort to the activity, which is also being watched closely all over the world. But governments have yet to demonstrate the political will to make the procedure effective.

Generating politics

All in all, what with assorted, perverse and often enormous subsidies to fossil fuels and nuclear power, and more modest but more visible subsidies for renewables; with inadequate accounting for risks; and with arbitrary and distorted provisions for externalities, only one conclusion can be drawn. As far as comparative costs are concerned, the choice of generation is political, not economic. Electricity costs stated as so many cents or pence per kilowatt-hour are just window-dressing after the fact, an artefact of prior decisions otherwise concealed. The same applies to the other original nineteenth-century criteria for choice of generation. Size and location are profoundly affected by politics, especially planning constraints on siting and operation. So is connection to networks. Once we acknowledge that the choice of generating technology, including its type, size, location and network connection, is fundamentally political, electricity policy takes on a significantly different flavour.

In particular the role of government appears in a new and disconcerting light. The headline purpose of liberalization was supposed to be to remove government from the business of making, selling and using electricity. That was always an exaggeration, to put it mildly. Regulation was and is construed as keeping government at arm's length; but this is only plausible when nothing too untoward happens. If the lights go out, the government is in the front line, no matter who else may be nominally responsible, as governments in California, Auckland, Sao Paulo, Ontario, the northeastern US and elsewhere have recently discovered with discomfort.

Since the oil price shock of 1973 governments have espoused what they call 'energy security' as a primary objective of energy policy. They have yet, however, to realize in practice just how different the 'security' issue is for electricity policy in particular. In this context, politicians usually construe 'energy security' or 'security of supply' to mean a secure supply of affordable fuel for electricity generation, especially oil and more recently also natural gas. For fuel-based electricity, governments in most OECD countries appear to believe that the main problem for 'security of supply' is now with imports of natural gas, potentially over long distances and from regions politically unstable. That is undoubtedly an issue, and important; but most attempts to address it nevertheless miss the much more immediate vulnerabilities closer to home.

As regards electricity, what matters is the security of supply of the services delivered by electricity. What matters is that the lights stay on. Interrupting fuel supply for fuel-based electricity does indeed pose a potential threat. But electricity is a process. Any interruption of the process, at any point in the system, may make the lights go out; and the process can be interrupted much more immediately and abruptly than by loss of fuel supply. Some such interruptions are local – a blown fuse, a burned-out incandescent lamp; you can rectify them readily. Others, however, may affect the entire system, in minutes or even seconds, blacking out a whole country. A storm bringing down a transmission line can cause a cascade collapse over a vast area, and leave it without electricity for days. In recent years such major blackouts have struck Brazil, Canada, France, India, Italy, Sweden and the US, among others, costing untold sums. Moreover, such a catastrophe may be triggered not only by an act of nature but also by an act of human malevolence. This possibility is already exercising governments at senior levels.

The part of the system most vulnerable to disruption is not generation but the network, including cables, towers and substations. Nevertheless, generation is the key to reducing the vulnerability of the network. The most effective measure is not to reinforce the network but to reduce its importance – in particular, to locate generators closer to loads, minimizing the amount of circuitry needed to complete the electricity process. If you are seriously concerned about the security of supply of electricity services, you may come to consider small-scale generation close to the loads a form of insurance worth paying a premium for, especially if you have sensitive loads. This will be yet more so for local infrastructure generation such as photovoltaics, requiring no fuel and not therefore vulnerable to disruption of fuel supply.

To the extent that governments consider themselves responsible for keeping the lights on, policies to enhance the security of supply of electricity services should therefore include measures to foster on-site infrastructure generation. Governments must recognize that real energy policy is much wider than traditional fuel and power policy, and that traditional methods of comparative costing of generation are grossly inadequate. Governments have ready access to appropriate policy levers, notably differential taxation of energy infrastructure assets, including buildings. To implement such policy, however, a government must have the courage to challenge the stultifying inertia of tradition – traditional concepts, traditional institutions and traditional mindsets.

Where will this happen, and when? Some look to the vast areas of rural developing countries such as China, India and much of Africa, in urgent need of electricity services and unencumbered by legacy assets and legacy traditions. The opportunity to leapfrog directly to innovative electricity systems and institutions is there for the taking. Unhappily, however, in precisely those parts of the world most in need of innovation, electricity policy tends to be in the grip of tradition more hidebound even than that in OECD countries. The implication is unmistakable. If the world is to gain the benefits of innovative electricity, starting with innovative generation, OECD countries will have to show the way.

Even in the most liberalized frameworks, OECD governments cannot sit back and pretend that electricity policy is now out of their hands – that private enterprise is henceforth responsible. The public does not believe it, because it is not true. For generating change, and for keeping the lights on, governments must forthwith take the lead. The time to do so is now.

7

NETWORKING CHANGE

As electricity systems worldwide battle through turbulent change, one key question emerges: what is the network for? Unfortunately, policy-makers rarely ask this question, much less answer it. That may be because they think they already know the answer, and see no reason to doubt it. It may, however, be because the question simply does not occur to them. It should. As electricity evolves, a central theme must be the evolving role of the network. Now and henceforth we need to ask and keep asking, explicitly, what the network is for. More precisely: we need to know what different parts of the network are for; who is involved; why; and how. The answers are changing as networks are changing. This evolution is going to go much farther than most of us yet realize, and perhaps much faster.

The traditional network is important as a starting point, not only because it still broadly prevails, but also because it continues to shape the thinking of planners and policymakers, even where traditional networks do not yet exist. Consider, then, how electricity networks came to arise. At its simplest, electricity does not need a network. In an electric torch or flashlight a single loop of wire connects the battery to the bulb – an electric circuit, closed and opened by a switch. This single circuit is an inherent feature of the process by which you produce electric light. It is either working or not working – on or off. However, as soon as you want to connect more than one load to a generator, or more than one generator to a load, the circuits multiply and interconnect: you have the beginnings of a network. In the early days of electricity, before the 1880s, arc-lighting systems might have a number of lamps running from a single generator. But the whole system – generator, wires, switches and lamps – all belonged to the same owner-operator, who paid the investment and running cost and used the entire system as desired. The

This chapter is adapted from 'Networking change', Working Paper 3 for *Keeping The Lights On*, Chatham House, 2004.

network involved was an integral part of the system, as essential as the crankshaft in a steam engine. Just as the output of a steam engine was motive power, so the output of an arc-light system was illumination. Moreover, just as you might own and operate the complete steam engine, not merely the crankshaft or the flywheel, so you owned and operated the complete arc-light system. You as owner-operator bore the whole responsibility for keeping those particular lights on.

At the beginning of the 1880s, however, this straightforward arrangement began to change. Thomas Edison, like several other contemporary entrepreneurs in North America and Europe, understood that only by pursuing economies of scale could he hope to make electric light competitive with gas light. With the generating technology then available, that meant using a larger steam engine and dynamo; but such a large generator would produce more electricity and therefore more electric light than a single client could then expect to use on one site. The obvious solution was to enrol multiple clients on multiple sites, interconnecting them all to the same large generator. That in turn entailed laying a network of cables between sites and through public space, a practice that might have been controversial but had already been sanctioned for gas pipes, not to mention water and drains. At the outset, nonetheless, as we noted earlier, Edison retained title to all the constituent parts of the system. His company owned generator, network, switches and lamps, just as if the whole system had been on one site. Edison charged his clients according to how many lamps they used; he was selling electric light.

In the mid-1880s, the invention of a practical electricity meter abruptly altered this arrangement. Once the meter intervened, the system was no longer delivering electric light. It was delivering electricity, bought and paid for by the measured unit and used in lamps the client purchased and owned. Soon thereafter, even the wires and switches on a customer's site belonged to the customer, not to the owner of the rest of the system. The network itself was divided between different owners. To be sure, the owner of the wires on a client's site, the client, paid no attention to them. So long as the lights stayed on, the wires were simply part of the building. If the wires malfunctioned, however, the client had to find and pay an electrician to fix them. If a lamp burned out the client had to buy and fit a replacement. Even as late as the 1920s some systems would replace lamps, just to keep customers using electricity; but in general as long as electricity reached the meter, the rest of the system bore no responsibility for keeping on the lights of any particular client. Within the 1890s the same became true of electric motors and thereafter of other electrical appliances on clients' premises. The responsibility of the owner-operator of the system stopped at an intermediate point on the network – the client's meter. The client became a customer for electricity, bought and paid for as a commodity with an established price per measured unit. The network acted as a delivery conduit, carrying electricity from the generator to the customer's meter, and dividing up the large output of the generator into quantities appropriately small for the aggregate load on the customer's side of the meter.

The electricity network became closely analogous to its precursor, the town-gas network – with one important difference. As we noted earlier, town gas was a commodity, a physical substance flowing through the network of pipes. It could be stored; the gas could be produced as desired and used as desired, with no necessary connection between production and use. Electricity, however, was not a physical substance, but a process, occurring instantaneously throughout the network of wires. Electricity, as a process, could not be stored; you had to produce it more or less exactly in the quantity and at the time that it was being used, continuously. This difference was to have a fundamental effect on the shape and operation of electricity systems, an effect now overdue for reassessment.

Subdividing the network

As Chapter 6 described earlier, from the mid-1880s the direct-current DC networks pioneered by Edison were joined and then overtaken by the alternating-current AC networks of Westinghouse and Tesla. A DC network with electricity flowing in one direction behaves much like a gas network. But an AC network behaves very differently. No gas analogue of AC electricity, surging rapidly back and forth, exists; AC is much more obviously a process than DC. An electric current always carries with it a magnetic field. If the current changes, so does the magnetic field, with profound effects on the combined 'electromagnetic' behaviour of the system. Designing and operating an AC system must take account of an array of continuous electromagnetic interactions far more complex than anything associated with DC. Indeed, a significant proportion of the energy associated with an AC system is required simply to maintain and manage the accompanying magnetic fields – so-called 'reactive power'. Yet AC, unambiguously a process throughout an entire system, made long-distance transport of electricity feasible. Paradoxically, the AC process, by making remote large-scale generation practical, facilitated treating electricity as a quasi-commodity for over a century.

From the 1890s onwards, the rise of AC brought about a fundamental change in the configuration and operation of electricity networks. As AC systems expanded, each AC network came to have two distinct subsections, one operating at substantially higher voltage than the other, linked by banks of large voltage-changing transformers in so-called 'substations'. The section at higher voltage carried electricity comparatively long distances with limited losses – so-called 'transmission'. The section at lower voltage delivered electricity to customers' meters – so-called 'distribution'. Eventually, both transmission and distribution became further stratified, with more and more levels of voltage for different purposes, interlinked by transformers, but all operating in real time as part of one complex continuous process. In such a system all the rotating generators on the system spin at the same rate, in step with one another; the entire system is 'synchronized'. An interconnected synchronized AC system is effectively a single machine, operating continuously in real time. In due course one gigantic AC machine might come to cover an entire country, or even more.

Initially, on any given system, the same owner-operator owned both the transmission and the distribution sections of the network; the only other owners involved were those on the customers' side of meters. The system owner-operator had to plan, install, operate, maintain and pay for the whole network, except the wires of customers. In some places, for instance London, different owner-operators established competing networks, often along the same streets. Elsewhere, however, the first network to be established in an area got such a head start that potential competitors sought more promising territory. In many places, for reasons of local politics, civil or municipal governments set up generators and networks. These authorities then used their political leverage, particularly over public space and waterways, to refuse permission to potential competitors to set up competing networks. This was the first manifestation of what subsequently became the almost universal practice, to grant an electricity network a political monopoly throughout a specified franchise area.

Promoters of the monopoly franchise, such as Samuel Insull of Commonwealth Edison in the US, argued that an electricity system with its network was a 'natural monopoly' – that permitting competing systems in the same locality would make electric light and motive power more expensive to users. The argument was controversial, and certainly self-serving. Whether or not the electricity network itself could then be considered a natural monopoly, other competing options included for instance on-site electricity generation, gas light and steam power. Nevertheless, natural or not, from the 1920s onwards electricity systems around the world, not just networks but whole systems, metamorphosed into political monopolies. As noted earlier, in the franchise area no one else could generate electricity to sell; no one else was allowed to run an alternative network through public space. Electricity users became captive customers. The monopoly franchise system operator, in turn, had to deliver all the electricity that users wanted, where and when they wanted it, at a price determined by government or a government-appointed regulator. Sometimes different parts of the network belonged to different owner-operators; for example, a number of low-voltage local distribution systems, separately owned and operated, might all receive electricity from a single high-voltage transmission system. But each separate part of the network was still a franchised monopoly in its area; and the entire interconnected network was still synchronized.

In this form of electricity system, what we have called 'traditional electricity', the role of the network was clear and unambiguous. The network was the conduit that delivered remotely-generated electricity to users. To fulfil the commitments of the monopoly franchise, the network had to have enough carrying capacity in all circuits to accommodate the peak load, with a margin of redundancy to spare. It had to reach all generators and all users on the system; adding generators or users meant adding network capacity. The network entailed investment and running cost, but did not itself produce revenue; expanding the network could be justified only if the consequent overall effect on the system would be to reduce the cost of generation or increase the income from users. In the centralized synchronized AC process, the network aggregated generation at any instant. That let the operator favour the output from the cheapest generators and 'dispatch' them accordingly, holding more expensive generators in reserve for peak loads. The network also offered a way to replace faulty generation, almost instantaneously if necessary to match load, provided the system had spare generation available and ready to produce.

Expanding and collapsing

For more than half a century, over much of the world, this arrangement worked remarkably well. Networks spread ever more widely, and became ever more interconnected, even across national borders. After the devastating disruption of the second world war, the expansion of electricity systems was unceasing; and all the systems were based on the same common technical model. Large power stations generated electricity as synchronized AC and delivered it to users over networks including high-voltage transmission and lower-voltage distribution. Pursuing economies of scale, system planners designed and built ever-larger stations, ever farther away from users, requiring ever-longer transmission networks at ever-higher voltages. The customers of monopoly franchise systems paid all the costs, but were generally unperturbed by their captive status, because until the early 1970s the cost of electric light, motive power and other electricity services steadily declined. The traditional technical configuration was so robust that it could function under many different institutional arrangements. Government owner-operators; private owner-operators; cooperatives; municipal, regional or national organizations; capitalist, communist or colonial regimes; all embraced the central-station synchronized AC form of electricity system – with admittedly varying degrees of success.

Even in OECD countries, the success of traditional electricity was not without controversy. Networks expanding into rural areas, financed by lavish subsidies from taxpayers, wiped out many local electricity systems based on wind or microhydro. Transmission towers on rural skylines attracted vehement objections even in the 1950s, when power stations themselves were still accepted without protest. However, objectors did not dispute the desirability and indeed necessity of transmission lines, as an inherent feature of the electricity system. They argued only to put a line somewhere else, 'not in my back yard', probably the earliest occurrence of NIMBY.

Perhaps the first serious setback to traditional electricity was the system collapse of November 1965 that blacked out most of the north-eastern US and eastern Canada. By that time people in most parts of most OECD countries took traditional electricity pretty much for granted, assuming that someone somewhere was keeping the lights on. The big blackout was a dismaying jolt. It demonstrated what most people had long since forgotten, that electricity is a process – that a traditional electricity system is a single gigantic machine operating in real time, and that it can also shut down in real time, over thousands of kilometres, in minutes if not seconds. The blackout prompted much breast-beating by electricity companies, who forthwith set up the North American Electric Reliability Council (NERC), charged to prevent any recurrence of such a problem. The members of NERC all held franchised monopolies. They proposed various measures, including enhanced redundancy on interconnected networks; regulators gave the go-ahead; and in due course the captive customers paid, as a form of compulsory insurance. Other OECD countries, whose electricity users could mostly afford the cost, took similar measures. Wherever possible, systems opted for multiple redundancy, both generation and networks, to address the problems of reliability. But the status of traditional electricity as a process occurring throughout a single machine meant that the possibility of shutdown remained inherent in the technical configuration of the system. In subsequent years, one system collapse after another, all over the world, underlined this inevitability. We don't always hear about them, but blackouts are long since endemic.

Dismembering monopolies

The year 1965, indeed, could be said to be a watershed for traditional electricity, the height of its success and the beginning of its decline. The mid-1960s saw the advent of rapidly expanding programmes of investment in nuclear generation, and also in other very large steam-cycle stations, many of which in due course experienced severe delays and cost overruns, some even cancelled when nearly complete. Nuclear stations provoked mounting opposition in many countries. Major hydroelectric dams in developing countries caused bitter controversy. Gaseous emissions, notably sulphur and nitrogen oxides, and other forms of pollution made fossil-fuelled electricity generation a key target of environmental concern, prompting legislation and controls. People still expected the lights to stay on; but electricity organizations, from having been unquestioned benefactors, gradually fell into public disfavour. Networks, especially the visual impact of transmission lines, still excited popular hostility; but from the 1970s onwards power stations themselves became the chief offenders.

In 1978 the US government under President Jimmy Carter enacted ground-breaking energy legislation, including the Public Utilities Regulatory Policy Act (PURPA). By fostering the concept of 'non-utility generators' permitted to sell electricity into existing systems, PURPA challenged the long-standing assumption that an electricity system had to be an integrated franchised monopoly. Practical implementation of PURPA was hotly controversial. According to PURPA, a 'non-utility generator', perhaps a wind farm or a cogenerator, was to be paid the 'avoided cost' that the system would otherwise incur. As you might expect, calculating this putative 'avoided cost' provoked bitter disputes and recriminations. PURPA nevertheless opened a whole new direction for the technical, financial and institutional evolution of electricity. Policymakers around the world followed its ramifications intently.

For a variety of reasons, therefore, by the late 1980s traditional electricity organizations were vulnerable to the sudden challenge presented by free-market advocates propounding a new vision for electricity. Electricity, they declared, should no longer be a 'utility' provided by a franchised monopoly, either a part of government or regulated by government. Electricity should be a commodity like any other, bought and sold in a competitive market just like oil or coal – or perhaps more like natural gas, because electricity, like natural gas, required a permanent network of conduits to deliver it. Chile under Augusto Pinochet launched the free-market concept of electricity. The UK government of Margaret Thatcher embraced it enthusiastically. In less than two years, from 1988 to 1990, the concept evolved frenetically from purely theoretical to practical reality. Within another year or two British electricity evangelists and fellow believers had spread the new gospel of 'privatization' and 'restructuring', or 'liberalization', across the world. But its hectic inception left fundamental questions not only unanswered but unasked, particularly about the role and function of the network in liberalized electricity.

Network as market

For nearly a century an electricity network anywhere in the world had fulfilled a concise and well-understood set of functions. Its primary role was simply to deliver electricity from generators to users, as a conduit analogous to a gas pipeline. Since almost all generators were thousands to many millions of times larger than almost all loads, the network divided the large output of individual generators into quantities small enough for individual loads. The network also allowed a central controller to select the cheapest generators to serve the total load on the system at any instant, and to replace a faulty generator with a backup generator, almost immediately if necessary to keep the lights on. The owner-operator decided how to maintain, modify or expand the network to fulfil these functions; and the captive customers of the monopoly paid whatever charges the government or regulator authorized. The arrangement was cogent and coherent, and at least in OECD countries it worked, not flawlessly but well enough to keep most lights on almost all the time. Where it worked less well, in communist or developing countries, the reasons were usually obvious, and political and financial rather than technical.

Electricity liberalization, however, involved a fundamental change in the status and function of the network. The guiding concept of liberalization was to create a 'market' in electricity. In principle, each generator would bid to sell its output, and each user would bid to buy the electricity required. Each transaction would entail agreeing to deliver and accept an appropriate amount of electricity over a stated period of time, at an agreed price per unit. The aggregate of all the electricity transactions in progress at a given moment would determine the amount of electricity flowing through each part of the network at that moment. In effect, the network itself would become a sort of 'marketplace' in which buyers and sellers would meet to do business. The more multifarious the network interconnections, the more liquid and effective the market. What mattered was not the actual execution of transactions but the very possibility of transactions.

To liberalize electricity systems, governments passed arrays of legislation establishing new frameworks for companies, markets and business relations. The accompanying rhetoric often referred to 'freeing the market' and removing government from the picture. According to the rhetoric, government would empower a regulator to launch the undertaking and guide it through its formative stages. The regulator would then recede gradually into the background. The market in its wisdom would demonstrate the most efficient and economic way to keep the lights on.

It did not work out quite like that. Free-market theoreticians came up against a dilemma. True, generators and users might participate in a competitive market; but they could do so only by means of a physical network that linked all participants, allowing them to share in the continuous electricity process. This physical network, far from being competitive, was still a monopoly, if anything stricter than before liberalization. Every market participant, whether buyer or seller, had to comply not only with technical protocols for access to and use of the monopoly network, but also with a whole new stratum of financial and institutional rules that the regulator propounded and enforced. These rules proved to evolve with dismaying speed, in ways that too often appeared incoherent and arbitrary. Instead of receding into the background, regulators expanded inexorably. Impinging on every transaction they became key players, with their own agenda, in what looked less like a free market and more like a free-for-all. In the UK, for instance, the regulator, after various vicissitudes, eventually became known as the Office of Gas and Electricity Markets, OFGEM. However, since its key responsibility was and is oversight of the monopoly networks, OFGEM might more accurately stand for Office of Gas and Electricity Monopolies. Similar considerations apply to regulators wherever electricity is being liberalized.

The theoretical neatness of the free-market idea collides with the reality of electricity as a process, in the existing traditional central-station synchronized AC system. In a complex multiply interconnected network, electric currents flow according to the laws of physics, not those of commerce. Keeping the system stable demands not only that total electricity generated must match total load as it varies, instantaneously and continuously, but also that other aspects of the process stay within acceptable bounds – a whole catalogue of so-called 'ancillary services'. For instance, the reactive power that maintains the accompanying magnetic fields in and around the network does not register on a conventional electricity meter; but without it the AC system cannot operate. Some versions of electricity market assumed that generators would provide reactive power as before, without additional payment; market designers made no provision for any suitable transaction. Other market designers found themselves having to provide for such essential ancillary services under duress, impromptu, ad hoc and expensive.

Those who liberalized electricity appeared to be proceeding on the premise that, despite the consequent sweeping institutional and operational changes, networks would continue to function technically more or less as before. But traditional networks were already under significant stress, which liberalization tended to aggravate. Apart from the remote but non-trivial possibility of system collapse inherent in traditional synchronized AC, reliability and power quality were already coming up the agenda as issues. Networks, once part of the solution, were becoming part of the problem. In OECD countries and indeed elsewhere an increasing proportion of loads are now said to be 'sensitive', requiring not only an uninterrupted flow of electricity but very stable voltage and frequency. These loads, such as paper mills and similar continuous process plants, microchip manufacturers, data-processing centres and server farms, also tend to be involved in activities with high added value, in which even brief outages or disturbances can be alarmingly expensive. A one-hour outage at a credit-card centre can incur costs in seven figures; a one-second outage on a chip line may ruin a multi-million dollar batch of chips. The desired standard is sometimes expressed as 'six nines' – that is, 99.9999 per cent reliability. Such a standard, however, is beyond the capability of a traditional synchronized AC network. Networks that used to smooth out and reduce disturbances are now more likely to cause them; and a synchronized AC system can propagate a disturbance a long way very fast.

Network innovation

Nor are sensitive loads the only problem for networks. At the beginning of the 1990s, liberalization coincided with the emergence of gas-turbine generation for baseload operation. As noted earlier, this marked a substantial break with tradition, in which a better power station was always a bigger power station farther away. A gas-turbine generator firing natural gas can be economic at a much smaller scale, as well as cleaner and more environmentally acceptable; you can therefore site it much closer to loads. That immediately has intriguing implications for networks. Other innovative generation adds further complications. Technologies already available or soon to be include wind farms, onshore and offshore; small cogeneration, using gas engines, Stirling engines, microturbines and fuel cells, even down to domestic scale; local wind generation, not only rural but also urban; local microhydro; and photovoltaics in many different applications. All these options require suitable network connections, operating protocols and other arrangements, often far removed from traditional.

As yet, provision of such network arrangements for innovative generation tends to be grudging, ad hoc and unpredictable. Network operators and regulators tend to cling to traditional mindsets about networks, regarding small-scale generation as at best uninteresting and not infrequently an active nuisance. But the pressure from innovative generators is steadily mounting; and efforts to develop appropriate protocols, regulation and legislation are under way in a number of countries. One consequence is already certain. As new generators and new users bring new attributes to electricity systems, the configuration and operation of networks is going to change. Networks that once simply delivered electrons may now have to make money, deliver transactions or deliver services, or perhaps all three. Networks may even foster – not hinder – change. But the transition will be neither smooth nor easy.

Network technology itself has lately seen major innovation emerge, albeit thus far more in theory than in practice. Even as liberalization came into the picture, substantial new breakthroughs were offering welcome additional capabilities to traditional synchronized AC networks. A suite of technologies called Flexible AC Transmission System or FACTS allows network operators much more subtle control over flows of electricity through the many different circuits of a high-voltage network. Power electronics, able to switch and modify electric currents more than a million times those in conventional electronics, gives network designers and operators impressive leverage over system behaviour. Even Edison's favourite, direct current, has reappeared as a strikingly attractive addition to network portfolios, in the form of high-voltage DC or HVDC. HVDC can double or triple the carrying capacity of high-voltage lines. It requires no synchronization of interconnections, and it can block the passage of disruptive transients.

In the US, for instance, the Electric Power Research Institute EPRI has developed a detailed 'road map' describing the successive introduction of innovative technologies, particularly network technologies such as these, into electricity systems. However, even as these technologies are becoming commercial, they have collided with an unfortunate corollary of liberalization. On many systems where FACTS technologies, for instance, would benefit operators and users, notably in the US, the necessary investment has not yet happened, because no one is quite sure who will pay for it, or how. Transmission operators lack incentives. Under prevailing US regulatory arrangements, for instance, even if operators are earning a fair rate of return, they have little incentive to invest in enhancements that are efficient but relatively low-cost. Investing in towers and lines provides a more capital-intensive solution and hence creates better earnings. In the UK, by contrast, the regulator, now OFGEM, has allowed network operators, both transmission and distribution, to invest substantial sums for maintenance and upgrades. The guiding presumptions, however, remain that the networks still function according to performance criteria laid down in 1977 under the monopoly Central Electricity Generating Board; that they are still a delivery conduit for electricity from remote large-scale generators; and that investment is directed essentially to this end.

Distributing generation

In recent years, to be sure, a succession of committees and other study panels have been examining the status and prospects for what is still often called 'embedded generation'. The adjective 'embedded' is ambiguous, but usually refers to generation connected to lower-voltage distribution sections of the network rather than to the traditional high-voltage transmission section. But 'embedded' also carries the unmistakable connotation of generation where it is not expected nor indeed supposed to be – generation in an inconvenient location. At one point the National Grid Company, operator of the transmission network in England and Wales, even declared that the rise of smaller-scale generation might endanger system stability, and that the central dispatchers might have to be given control over generators down to 10MW of output. In outraged rejection of this idea, smaller generators pointed out that the real stability problem rested with large steam-cycle units that could trip and send a 500MW transient hundreds of kilometres. Small generators, comparable in size to loads, could have no such drastic impact. The stability argument seems to have faded into the background;

but the traditional mindset persists. Some traditionalists continue, for example, to insist that all so-called 'intermittent' generation, such as wind, should have to provide full-capacity 'backup' from dispatchable fuel-based generation. Despite such impediments, as smaller-scale generators gradually establish themselves, the description 'embedded' is being supplanted by 'distributed'. Until recently, however, the changes in network arrangements to accommodate distributed generation have been mostly marginal and tentative; they still frustrate and discourage prospective developers of innovative generation of whatever kind.

A more promising approach, once again pioneered in the US, adopts the catchphrase of computer users, 'plug and play'. Computer hardware said to be 'plug and play', as the phrase indicates, conforms to a standard technical protocol. You can connect it to the rest of the system with no further specific preparation. It will forthwith function as desired, as a part of the system. In the case of electrical hardware, loads such as lamps, motors and electronics have long been 'plug and play'. So long as a device meets the requisite safety and other standards, you can connect it to a traditional synchronized AC network, indeed plug it into a socket, and turn it on immediately. Networks, however, have never accepted generators on the same basis. Under the traditional monopoly franchise, you might well have, say, a standby diesel generator for emergencies; but you could use it only in isolation, with your on-site system disconnected from the network. If you wanted a generator to operate continuously while connected to the network, it had to satisfy not only the requisite electrical-engineering protocols for basic safety and performance, but also operational protocols laid down by the network operator, restricting its use. In traditional electricity the system did all in its considerable power to obstruct, and if possible prevent, on-site generation and cogeneration. After liberalization, this obstructiveness was supposed to give way to competition between generators on a level playing-field. But networks have yet to welcome small-scale and on-site generation with anything but severely modified rapture. They have had little if any incentive to do otherwise.

Encouraging signs are nevertheless perceptible. In the UK, for instance, an 'Embedded Generation Working Group' reported in 2001, and its key recommendation led to the establishment of a 'Distributed Generation Coordinating Group'; the change of designation from 'embedded' to 'distributed' was itself noteworthy. This group, including high-level participants from large and small generators, network owners and operators, users, regulators, consultants and government, in turn reported three years later. It continues consultations on a forward programme of further efforts to eliminate barriers to distributed generation, and indeed to provide incentives to distribution network operators to foster it. The detailed analysis undertaken by and for the group is impressively comprehensive, addressing a wide spectrum of issues affecting the status and prospects for innovative generation connected to networks at lower voltages and closer to users. Pending the establishment of the promised incentives to network operators, the achievements can mainly be characterized as clearing the decks, removing actual obstructions; but that in itself is an essential preliminary. True 'plug and play' for small-scale generators – which EU discussions call 'fit and inform', requiring no prior permission from the distribution network operator – is farther down the line. The group noted that 'Significant increases in distributed generation will fundamentally affect the design of distribution networks.' For 'design' you could read 'design and operation', or simply 'function'. The rise of small-scale local generation, and of optimized local systems closely linking generation and loads, will interact profoundly with the neighbouring network.

The UK exercise is also valuable and instructive in that it brings together network regulators with all those who use the network and rely on it, an opportunity to pursue in depth the question that opened this chapter: what is the network for? What should it be for as electricity evolves, and how should regulation help the network to evolve appropriately? In the longer term, what will the network look like, and how will it work? At the moment, for instance, 'transmission' and 'distribution' are useful descriptions of distinct subsections of the network, with distinct ownership, financial arrangements and regulatory frameworks. Will that continue? To all these questions we can decide the answers. None are predetermined. If policymakers, especially regulators, weigh long-term network options with open minds and imagination, change could accelerate dramatically.

Innovating finance

For obvious reasons, changing the technical configuration of the network towards one more amenable to decentralized generation will necessarily be gradual and protracted. It will only happen if financial arrangements and institutional frameworks also change. Indeed such conceptual and procedural changes are a critical prerequisite for technical change, for actual physical reorganization of network hardware. The financial arrangements that now underpin liberalized electricity continue to focus on the flow of electricity through meters and other measuring instruments in the network. They involve transactions based on this flow and on the price per unit of electricity as it flows – commodity transactions and a commodity price. Electricity finance continues to treat decentralized small-scale and local generators as equivalent to remote large-scale generators. From this viewpoint, what matters is the flow of electricity delivered into the network as a whole, not the electricity delivered to local loads with minimal or zero use of network circuits beyond the local meter.

However, if decentralized generation is to achieve its real potential, it will often become part of an integrated local system, optimizing generation and loads together for maximum performance, not on a short-term commodity basis but as an overall investment in system assets to deliver desired services. The flows of electricity into and out of this local system may not be trivial, but they will be of secondary importance to the overall financial status of the local system. The advantages of such an integrated local system are manifold. Designers, operators and users of this local system would have direct and immediate interest in combining and integrating system assets, especially loads, to get the best available performance, for reasons of economics, reliability and security, to keep the local lights on. The cumulative effect of such integrated local systems would be a dramatic increase in the effectiveness of the energy infrastructure, reducing both dependence on fuel and vulnerability to disruption.

One early move to support a transition in this direction would be to broaden the financial relationship between the generator and the wider network. We could introduce this change almost immediately, and with ample precedent. In traditional electricity, users with loads have long paid a two-part or 'binary' tariff, described as a fixed charge for the connection plus a variable charge for the amount of electricity used. In effect the user is paying the fixed charge in order to have the system assets, in particular the network, available when required. A financial arrangement of this kind, taken for granted for loads, could well be extended to apply to generators, especially if the generators are broadly comparable in size to loads.

In recent years some companies owning and operating large generating stations in competitive markets such as the UK have lobbied vigorously for 'capacity payments' for keeping generating assets available, whether or not they are actually delivering electricity at any given time. Regulators argue that such fixed payments unconnected to flows of electricity would undermine the electricity market, in which transactions all depend on trading quantified units of electricity at a quantified price. But owner-operators of traditional power stations point out that they get paid only when the station is delivering electricity. Station output is not like a genuine commodity such as wheat, oil or pork bellies. They cannot store it until they can sell it at an acceptable price. The nature of the market is such that unit electricity prices may be too low even to cover the owner-operators' cost of capital, which they have to pay whether or not stations are generating. In consequence a lengthening list of major UK power stations have been in acute financial distress. Some foreign owners have simply walked away, leaving stations in the hands of creditors. In any case, the price you pay as an electricity user is determined in substantial part not by the market but by the charges for use of the network. The regulator sets these charges. The users who pay them remain in this respect captive customers, exactly as before liberalization. In such a context absolute insistence on a pure commodity market in electricity is ideological, not practical. It appears less and less likely to deliver stable or reliable long-term electricity services.

Low prices or low bills?

A key aspect of the problem may be simply the long-standing conviction that the aim of liberalization and its attendant regulation is to achieve the lowest possible unit price for electricity. Regulators have taken that objective as self-evident, and congratulated themselves on progress in that direction. Large users, especially electricity-intensive industries, have broadly agreed. But most electricity users, including households and small businesses, appear to have little idea of the unit price of electricity, nor do they much care. What matters to them, what they want, is not a lower unit price but a lower bill. The one does not necessarily imply the other. Indeed, as noted in Chapter 6, intriguing recent research suggests that a higher unit price may actually lead to a lower bill, by encouraging upgrading of end-use equipment. At the very least, regulators should reappraise the underlying objective, and ask whether a low unit price is either necessary or even desirable, given its corollary implications for generators and users alike, in terms of reliability, security, efficiency and environmental impact.

Questioning the desirability of a low unit electricity price does not imply questioning the desirability of a market. What we need to question is not the idea of a market, but the nature of this market – what is being bought and sold, by whom and how. Since liberalization, we have modelled the electricity market mainly on the market for natural gas; but we need not. We need not confine suitable transactions to trading in defined quantities of electricity over defined time periods at agreed prices. We could readily redraft contracts to include, for instance, payments for availability of assets, or to reserve capacity of assets, at fixed prices with no reference to flows of electricity, if both counterparties so desired, provided the regulator agreed. Nor need counterparties be, for instance, a generator and a user, as would normally be the case for a commodity transaction. Indeed network owner-operators might become active participants. In appropriate circumstances, moreover, an asset-based contract could also include network assets. That would become yet more important with the significant emergence of private wires as important circuits within the wider network. Such possibilities also suggest a wider and more creative role for the regulator. In particular the regulator could help to shape contractual agreements, to identify and incorporate appropriate incentives to foster innovative electricity in all its varied manifestations.

For that to happen, the regulator must have the requisite authority and backing from the relevant government. That in turn means that the government must understand the true range of technical, financial and institutional options and choices, both traditional and innovative, available to electricity policy. Government will have its own objectives for electricity: first and foremost to keep the lights on, to do so reliably over time at acceptable cost and with acceptable environmental impact. Government will have to decide how best to achieve these potentially competing objectives. In a liberalized context, however, except when it acts on its own behalf as an electricity user, it cannot directly implement electricity policy, in the sense of investment or pricing. Instead it must set the legal and regulatory framework, and wait to see whether other players act appropriately. In the present transition phase beyond traditional electricity, government still has significant leverage, through taxation, grants and other financial measures, standards for performance of assets including buildings, and its own procurement policy. It can also give guidance to the regulator, particularly over networks that remain more or less unbroken political monopolies. But governments in general tend to leave such specialized matters to the regulator, requiring only the regulator's assurance that the lights will stay on. As and when the regulator cannot give such assurance, or proves mistaken, governments will face a severe challenge. They would do well to study the electricity options – all the options, innovative as well as traditional, accurately evaluated and compared – beforehand, when they can act rather than react.

Sustainable networks

As they do, they might well raise their sights beyond the next election. Governments genuinely desiring the most reliable, economic and sustainable electricity services for their citizens in the long term now have an unparalleled opportunity within their grasp. All over the world, governments have endorsed the global aspiration to sustainable development. Many have also committed themselves to reduce emissions of greenhouse gases, particularly carbon dioxide from fossil fuels. A rapidly expanding body of evidence indicates that a key dimension of sustainable development will be sustainable energy; and the centrepiece of sustainable energy will be sustainable electricity.

Sustainable electricity will require a 'sustainable network'. 'Network' itself now overstates the case; 'networks' would be more accurate. The system is no longer a single synchronized machine operating in real time, but a much more loosely interconnected and interoperating array of significantly independent sub-networks, each with both generation and loads. The networks still produce significant external benefits that grow with size. They may not need to deliver electricity thousands of kilometres; but the possibility to do so if necessary has beneficial effects on local pricing. They also help to smooth and diversify the effects of fluctuating generation as well as fluctuating loads.

Early steps in this direction look promising. Work is already under way to devise ways to take advantage of the flexibility of insensitive loads such as refrigerators, freezers and air-conditioners. In the transformed future, instead of 'load-following', systems suitably equipped can 'load-match' by adjusting loads to correspond to generation, not merely vice versa, as appropriate. That may be easier to manage in local systems, but will depend on the particular mix of loads and generators. At least two UK entrepreneurs, for instance, are advocating the inclusion of simple circuitry in freezers and refrigerators, that can sense the frequency of the alternating current on the network. When system load exceeds generation, the frequency falls. Freezers sense the fall, and, on a randomized basis, disconnect from the system temporarily, relieving the strain on the system. Conversely, if generation exceeds total load, disconnected freezers reconnect, once again improving the match between generation and loads. A freezer is effectively a 'thermal store' on the system, acting as a buffer to smooth out system changes, while continuing to keep its contents well frozen. In effect, in addition to load-following generation, the system also has 'generation-following loads'. The necessary circuitry in the appliance would add only a few pounds at most to its cost of several hundreds. Governments so minded could stipulate that all suitable appliances offered for sale would have to include the relevant control circuitry, just as they stipulate other performance standards for electrical equipment.

Within sub-networks, circuits may carry synchronized AC or low-voltage DC as loads require and generators produce, with interconnections using electronics or power electronics as necessary. Between sub-networks, connections may be synchronized AC up to and including high voltage; back-to-back AC–DC–AC; or high-voltage DC as appropriate, again with interconnections using power electronics, including digital transformers, as necessary. The overall shape of the network, its 'topology' or connectedness, will be a mesh, with many nodes and many pathways between them – a sort of 'electricity internet'. Control of voltages and currents through and between circuits will be autonomous, with sensors, instrumentation and switchgear interlinked by telecoms and real-time online processing.

In this transformed electric future, designating some circuits as 'transmission' and others as 'distribution' will no longer be useful. Network assets, loads and generators may all belong to a single owner, or different owners in various groupings, including for example property-owners. The traditional monopoly franchise will be attenuated or completely eliminated, although we may still require some form of regulatory oversight, to enforce protocols and transactions. We may still mediate some transactions by meters and measured flows of electricity.

Others will be longer-term contractual relationships, such as agreements for availability or reserve of assets, or for services, of a specified quality over a specified period. Parties may agree on fixed or variable payments with no necessary reference to flows of electricity. Investing in relevant assets will depend on making suitable financial arrangements for subsequent revenue from using the assets. Decades hence, when fuel-based generation has dwindled and infrastructure generation ramified, electricity may gradually become not only physically but economically invisible, as yet another function of infrastructure. It will be paid for not by the measured unit but by investing in infrastructure assets. In many circuits the flows of electricity will not be measured at all, because no transaction depends on them, any more than in Edison's day.

Networks are best positioned to lead this transformation. They should foster change, not merely follow it reluctantly. The evolution of networks should trace the same arc as generation: initially to higher voltages, longer distances and more centralization, now back towards lower voltages, shorter distances and decentralization, including integrated optimized local systems. Sustainable networks are the key to sustainable electricity.

8

DECENTRALIZING NETWORKS

Decentralized energy is nothing unusual. Nature is not centralized. Natural energy is everywhere, in sunlight, wind, water, plants and animals. It runs the planet. However, we take all that decentralized energy pretty much for granted. What we notice is the centralized energy we ourselves distribute. We extract coal, oil and natural gas from concentrated central sources – coal seams, oilfields and gasfields. We then move it from place to place – distribute it – in mobile transport such as ships, trains and trucks, and in infrastructure networks such as pipelines. We use energy from fuel where and when we wish, converting it into more useful forms such as heat, light, sound, and kinetic energy of movement. We likewise gather, convert and distribute some natural energy flows, notably those of water, wind and sunlight. To distribute both fuel and natural energy flows we also convert them into one particularly versatile form of energy – electricity.

Like natural gas, electricity requires an infrastructure network. Unlike natural gas, however, as we noted earlier, electricity is not a physical substance, not a fuel nor a commodity. It is a process, happening simultaneously throughout the whole system infrastructure – generators, network and loads. Indeed without the infrastructure electricity does not even exist. We don't actually want electricity itself. But we can convert it in turn into all the forms of useful energy, easily, cleanly and conveniently. Furthermore we can generate electricity anywhere, over a vast range of scales, from watch batteries to turboalternators, in almost any quantity from imperceptible to overwhelming; and we do.

We pay, however, particular attention to one form of electricity, in which large central stations generate synchronized alternating current and send it out to users over a network that includes long high-voltage

This chapter is adapted from 'Decentralizing networks', *Cogeneration and On-Site Power Production*, January–February 2005.

transmission lines. Since the 1880s, electricity systems based on this common technical model have spread all over the world, bringing electric light and motive power and other benefits on which modern society now depends. Large-scale centralized generation of electricity has become so important, and so dominates our thinking, that we have long tended to discount the many alternative forms of electricity generation that are smaller in scale and less centralized. In recent years, however, these forms of generation have become harder to overlook. Based, for instance, as we have noted, on wind turbines, microhydro, diesel engines, gas engines, Stirling engines, microturbines, fuel cells and solar photovoltaics, they tend to come in unit sizes much smaller than central-station generators, usually less than 5MW. Since individual units or clusters of units may be widely dispersed across an electricity system, rather than being centralized, these technologies have come to be called 'distributed generation', an increasingly promising form of decentralized energy.

Although traditional electricity generation is centralized, the loads that use the electricity, such as lamps, motors, heaters, chillers, and electronics, have always been widely distributed and dispersed. Except for the very largest loads, such as pot-lines in aluminium smelters, loads are thousands to millions of times smaller than central-station generators. This mismatch in scale between generation and loads requires the network to divide up the large output of a generator into flows appropriate to the loads – that is, to distribute the electricity. Alternatively, of course, generation itself could be distributed, closer to loads in both location and scale.

The reason why it is not is historical, and ripe for change. To recapitulate and summarize: in the early decades of electricity, generating technologies were based on water power and steam power. The economies of unit scale of steam engines and turbines, water turbines and alternators meant that a bigger generator produced cheaper electricity. That was the premise on which Edison and his competitors set up the first central-station systems. The savings on investment in larger generators more than made up for the extra investment in the necessary network. In the subsequent century this premise continued to prevail, up to generators of gigantic size and networks to match, entailing likewise gigantic investments. The investments were possible because the monopoly franchise made captive customers carry the risks, which by the 1980s sometimes proved equally gigantic. Nevertheless the arrangement made electric light and other electric services available and affordable over much of the world.

It was so successful that by the end of the 1980s free-market enthusiasts decided that electricity, too, was a commodity that should be bought and sold in a marketplace. In a rapidly expanding list of countries they abolished the monopoly franchise, broke up traditional integrated systems, separated generation from networks and made generators compete to sell their output to users. One of the many unexpected consequences of this 'electricity liberalization' was to make distributed generation look distinctly more promising. Introducing competition made investment in traditional large-scale generation much riskier; and abolishing the monopoly franchise transferred the risk of investment from captive electricity users to skittish shareholders and bankers. At the same time technical innovation widened the range of generating options. Cheap and abundant natural gas made gas-turbine generation the new favourite, breaking at last with the long presumption that a better power station was always a bigger one farther away. Gas-turbine generation could be at once cheaper, cleaner, more efficient and closer to users. Other even smaller generating technologies, some likewise fuelled by natural gas and others based on renewable energy, also began to attract attention. Compared with traditional generation they were easier to site, quicker to build and commission, and much cleaner. But they still faced problems.

Starting from here

Some arose from existing networks. Traditional electricity regards the network as a 'natural monopoly'. Natural or not, it has long been a political monopoly almost everywhere. Its essential configuration is radial, from the centre outward. One-way flows carry electricity from large-scale remote central generation along high-voltage transmission and lower-voltage distribution lines to dispersed users. This radial one-way configuration is less appropriate, however, for distributed generation, smaller and closer to users, often most usefully connected at lower voltages. Distributed generation has more in common in scale and in attributes with loads than it has with centralized generation. Connecting a 500kW microturbine has much the same effect on the system as disconnecting a 500kW motor. But generation connected at low voltage may cause current to flow in the opposite direction through the neighbouring circuits, confusing protective devices and potentially endangering maintenance staff. On the other hand, such local generation may provide voltage support and reduce the need to reinforce the network itself. Such trade-offs are now under intensive consideration by electrical engineers and system planners. The ideal arrangement would be technical protocols such as those for loads. If your local generator complied with the protocol, you could then connect it just as you do loads, effectively by plugging it in and turning it on. But such convenient arrangements are still mostly under negotiation in Europe, North America and elsewhere. One point of dispute is the usual one: who is to pay for the requisite reconfiguration of networks?

Before liberalization, network investment and running cost tended to be aggregated with those of generation, and paid for by the aggregate revenue from users, as mediated by government or regulator. After liberalization, governments and regulators expected the network to function also as a market-place linking sellers and buyers of electricity. In other respects, however, they expected it to operate as before, and with the same configuration. In the new market framework, the regulator would impose charges for using the network to carry electricity between buyers and sellers. In effect, despite liberalization, the network would continue to be a regulated monopoly. In practice, despite the rhetoric of free-market enthusiasts, close to half the price of a unit of electricity was thus determined not by a market but by regulatory decree. It still is.

Some policy people nevertheless cite the purported cost of a unit of electricity from different generating technologies, often in fractions of a penny per unit, to claim, for instance, that large-scale remote fossil-fired generation is 'cheaper' than smaller-scale renewable or cogeneration closer to loads. With no qualification as to the accounting or financial framework, tax treatment, subsidies, risks, system and network effects or other essentials, such cost comparisons are meaningless. They should have no influence whatever on policy. Policy determines costs – not the other way round.

This further underlines a crucial point about electricity. You can generate electricity without fuel, but not without infrastructure. Electricity depends absolutely on an infrastructure of physical assets. However, by treating electricity as a commodity, the 'electricity market' makes the price of an ephemeral unit of electricity the determinant of all the financial relationships involved, including – crucially – investment. The revenue paid to a generator depends on the number of units sold and the price per unit. That in turn depends on whether the generator can connect to the system – be 'dispatched' by the network operator. For distributed generation of many kinds this is a serious constraint. A wind turbine generates when the wind is blowing, not when a dispatcher invites it to. A cogenerator responds to requirement for heat, not for electricity. Distributed generators are penalized for not being dispatched. But no fundamental law of electricity says that distributed loads should always be independent, while distributed generators, often of much the same size, have to respond accordingly. The problem of network stability arises not because a wind turbine or cogenerator fails to deliver a few megawatts; what triggers instability and causes blackouts is much more likely to be the abrupt loss of a traditional 500MW unit, or of the network circuit carrying its output. Replacing centralized with diverse and dispersed distributed generation on a system enhances rather than undermines its stability. Far from penalizing them, we should pay distributed generators extra, for the stability insurance they provide.

For these and other reasons, including environmental issues, planners and policymakers in both OECD and non-OECD countries are at last giving much closer attention to the potential for decentralized energy technologies such as distributed generation. In particular they have begun to re-examine the interactions between distributed generation and electricity networks. Their efforts to date are undoubtedly helping to foster the expansion of distributed generation. But they have yet to overcome a key problem. Existing networks, their configuration and mode of operation, came into being as a necessary complement to central-station generation. If we were starting now to establish an electricity system based not on central generation but on distributed generation, it would require a very different network – different in configuration, function and operation.

How decentralize?

Consider two situations, one with and one without an existing traditional network. Remember that traditional electricity, for all its historical success, has failed to reach two billion people. It may even be losing ground, as population outstrips expansion of traditional systems. Much of the world is indeed still waiting to establish electricity systems. The contrast between the two situations is straightforward. An existing network represents what has come to be called 'legacy' technology, already in place and operating. Any change will be constrained by the need to keep the system operating through the change, to keep the lights on. It also implies legacy institutions and a legacy mindset, committed to a certain way of thinking, acting and interacting – assuming, for instance, the primacy of centralized electricity. Where no network now exists, these constraints are absent. These parts of the world, however, usually have either limited competence or a tendency to aspire to the traditional central-station model, even when a decentralized alternative might be more effective. Moreover, because electricity infrastructure represents major investment and employment, it also brings with it significant political power, a potent centralizing factor.

In both situations, therefore, realizing the potential of decentralized energy will require positive policy measures to overcome these obstacles. Consider, first, an OECD country with a highly developed existing network. Why might it benefit from more decentralized energy, and what measures would foster this? In recent years one issue above all has come to dominate electricity thinking in OECD countries, the issue of reliability. Spectacular blackouts in wealthy neighbourhoods grab attention. People demand that something be done, and that governments do it. But people and governments alike have yet to realize that a traditional synchronized AC electricity system is in effect a single giant machine, extending perhaps for thousands of kilometres. It is operating in real time; and like any other machine it can also shut down in real time. The possibility is inherent in the configuration and operation of a centralized system. No amount of hand-wringing can change this. The obvious remedy is therefore to loosen the centralization, to reduce the interaction between widespread parts of the system. Distributed generation is the key.

Those with sensitive loads are already making the initial moves, to gain control of their own electricity and to keep their own lights on. But governments and regulators can accelerate the process. Governments can recognize that electricity is an infrastructure issue. Their most appropriate tax leverage is tax treatment of system assets – not only generation and networks but also, and most importantly, loads. Favourable tax treatment for integrated, optimized local systems, possibly including cogeneration delivering both electricity and heat, will give a powerful boost to the requisite reconfiguration. Government procurement for its own buildings and other facilities can set an example. That will also prime the pump for energy service companies able to deliver the complete package, installing, operating and maintaining local systems on the basis of contracts for services. Regulators can recognize that the radial one-way configuration of networks must evolve, into a meshed two-way network, with the requisite technical protocols. Private wires, as adjuncts of local generation, can show the way.

Regulators can acknowledge, belatedly, that infrastructure is paramount – that treating electricity purely as a commodity cannot deliver investment, reliability or stable business relationships. Regulators need to rethink the nature of transactions. They can incorporate and endorse appropriate payments for generation assets, for availability, for access and for use, just as they already do for networks. That will greatly enhance the attraction of small-scale generation, both renewables and cogeneration. Increasing the proportion of electricity so provided will also of course produce a corresponding reduction in carbon emissions, not by coercion or trade-off but as a welcome corollary of other benefits.

Successful implementation of measures such as these in OECD countries will greatly improve the likelihood that they will also be adopted elsewhere. Given the comparative freedom from legacy constraints, localities around the world still eager for electricity services will find that decentralized systems can be established even more rapidly and effectively, using small-scale technologies and local resources, under local control and locally financed, perhaps by micro-credit. Many examples already exist. Appropriate commitment by international agencies and technology suppliers can dramatically expand these activities. Those with outdated systems and those with no systems at all now have the opportunity for genuinely fruitful and mutually beneficial collaboration.

Who knows? Over time, decentralized energy might even become the norm, in human society as it is in nature.

9

GETTING THE STORY RIGHT

Electricity needs a new story. We've been telling the same old story for more than a century now. It's out of date, it's boring and it's wrong. If electricity is to meet the challenge it now faces, and seize the opportunities now to hand, electricity needs a different story, a better story, one that captures the dramatic changes now unfolding around the world.

You now know the old story pretty well. It's the one we've been telling all our lives. It's the traditional story about how we make and use electricity, about how we think of electricity, about how electricity fits into our lives, our economies and our societies. Those of us actively involved in electricity policy know the traditional story more or less by heart; indeed we are among the key storytellers. We tell it to politicians, financiers, business people, media people and electricity users, over and over. They listen to us. Sometimes they even act accordingly. The story we're telling matters. It matters particularly when the story we're telling is wrong.

In this traditional story, electricity is a fuel like any other fuel. In this story, someone makes electricity in power stations, and delivers it to users over a network of cables. A better power station is usually a bigger power station, farther away. Producers and users sell and buy electricity as a commodity, by the unit. Someone has to build and operate the power stations and the network. Someone has to finance these undertakings. Since the network is a 'natural monopoly', the government imposes regulations to ensure that all participants are treated fairly. Electricity users are independent. They buy their own electrical equipment and attach them to the system as loads. As they switch loads on and off, the rest of the system has to respond accord-

This chapter is adapted from 'Getting the story right', keynote address at 'Energy, Sustainability and Integration', the CCGES Transatlantic Energy Conference, York University, Toronto, 9–10 September 2005.

ingly. Someone has to keep the system stable from moment to moment. Someone has to ensure that the system has enough generation and network capacity to meet the maximum possible load that users can connect. Someone has to keep your lights on – someone else, not you.

For most of the past century this was a pretty good story, good enough to be told and retold all over the world. It helped to make electricity essential to what we think of as modern society. But the traditional story no longer makes sense. It is riddled with holes that are growing harder and harder to ignore. It has lost the plot.

The electricity story used to be a documentary, based on fact. Now it looks more like fantasy, wishful thinking, out of touch with reality. The reality today is that one-third of us don't have electricity at all. Those that do have trouble keeping the lights on. The International Energy Agency estimates that electricity will require investment of $10 trillion by 2030 – more one thousand billion dollars every three years. But the past decade has cost many in the electricity business their jobs, their shirts or their companies – hundreds of billions of dollars of losses already. Future electricity investment could now be so risky it might not happen. The main technologies of traditional electricity – large dams, coal-fired and nuclear power stations, and overhead transmission lines – are long since in major trouble, financial and environmental. Yet traditional electricity diehards are now trying to stampede us into more of the same, to make matters worse. The electricity story could become a horror story. We need to rewrite it, urgently.

To get the story right we have to get the premises right. To start with, we have to get electricity itself right. Reiterate the essentials; we should now take them for granted. Electricity is not a fuel. It's not a commodity. It's a process, occurring simultaneously and instantaneously throughout an entire interconnected circuit. A process cannot be stored. A fuel such as coal, oil or gas comes out of a hole in the ground at a particular place. If you want to use it anywhere else you must carry it there. But you can start the electricity process anywhere, in an extraordinary range of ways, from vast to minute. Electricity exists only in the infrastructure of assets that generate, deliver and use it, and through which it flows. Electricity is a function of infrastructure. Understanding this is the key to the necessary changes. You can produce and use electricity without fuel, but not without infrastructure. The flow of electricity through the infrastructure is easy to measure; but the price of a unit of electricity is ultimately arbitrary. The so-called 'electricity market' is illusory. The price of a unit depends not only on the price of any fuel involved, but on asset accounting, taxation, regulation, risk, subsidies, network and system effects, and other factors usually unmentioned. The arbitrary price of an ephemeral kilowatt-hour is not an adequate basis for the requisite finances, transactions and business relations.

Electricity means infrastructure

Instead of a quasi-commodity market, we should deal explicitly with the physical assets of the system – generators, networks and end-use technologies. What matters is this infrastructure – who owns it, who has access to it, who uses it and on what basis. What we need is not batch transactions in a quasi-commodity, but contracts for services.

To deliver electricity services more reliably and sustainably we need not only to upgrade the electric infrastructure, all of it, especially the end-use technologies, but to transform it. This is where the new story starts. Traditional electricity is based on a technical model dating back more than a century, to when the best available generating technologies were based on water power and steam power. Economies of scale in generating with these technologies shaped the model. As a result, all over the world, a century later, we still generate electricity in large remote central stations as synchronized alternating current, and deliver it to users over a network including long high-voltage transmission lines. The network is essentially radial and one-way. It also has to divide up the electricity to distribute it to loads mostly thousands to millions of times smaller than the generators.

This centralized configuration used to make sense. It no longer does. Consider some of the obvious drawbacks. Most central-station generators operate either intermittently or at only partial load most of the time, misusing costly assets. Fuel-based central generators waste two-thirds of the fuel energy before it even leaves the power plant. On many systems line losses cost another significant fraction. The configuration is inherently vulnerable to disruption, by mishap or malfeasance, over a wide area and almost instantaneously. Traditional electricity assumes that every load is essentially equivalent, requiring the same high quality of electricity. This is akin to our absurd water-management policy, in which we purify water centrally to drinking-water quality, and then use most of it for flushing toilets, washing cars and watering lawns.

In the same way, we produce high-quality electricity as required by sensitive loads, then use much of it for undemanding services such as heating and cooling. Most electrical loads, moreover, are inherently intermittent or variable; but large fuel-based generators are inherently inflexible. Traditional arrangements are almost a total mismatch.

Anyone who looks dispassionately at traditional electricity has to think 'We must find a better way to do this.' Fortunately, now we can. As we have seen in earlier chapters, the catalogue of innovative generating technologies already or soon to be available extends far beyond steam and water power, with very different attributes. Wind turbines, microhydro, biomass generators, photovoltaics, gas engines, microturbines, fuel cells, Stirling engines and microcogeneration all exhibit economies not of unit scale but of series manufacture. The more we make the cheaper they get. We can often locate these small-scale generators close to loads, even on site, dramatically altering network requirements and operation. Instead of a radial one-way network, a decentralized system would have a two-way meshed network, with loads and generators of broadly comparable sizes more or less uniformly distributed across the system. Monitoring and control technologies now indeed offer the possibility of completely self-stabilizing systems, in which loads and generators talk to each other continuously and react accordingly.

By moving towards innovative decentralized electricity we can tackle directly the most serious shortcomings of traditional centralized electricity. But first we have to explain and clarify this new story to policymakers. Ignore the reported 'cost of generation' by different means. It usually claims that traditional large-scale remote fossil-fired generation is 'cheaper' than smaller-scale renewable or cogeneration closer to loads. As noted earlier, such comparisons, in fractions of a penny per unit, with no qualification as to the accounting or financial framework, tax treatment, subsidies, risks, system and network effects or other essentials, including environmental effects, are meaningless. They should have no influence whatever on policy. The corollary bears repeating until it is second nature. *Policy determines costs* – not the other way round. That indeed should be the aim of electricity policy, sensibly and coherently developed.

The whole story, the whole system

Consider for example taxation. If we treat electricity as a commodity, taxation applies only to the unit price, and to batch transactions in measured amounts. Even for fuel-based electricity this is unsatisfactory. For what we might call 'infrastructure electricity', such as wind, hydro or solar, such tax treatment misses the point completely. If, however, electricity is treated correctly, as an infrastructure issue, tax policy should focus not on flows of electric current but on taxation of assets in electricity infrastructure – all assets, explicitly including end-use technologies and the buildings that contain them. To upgrade electric infrastructure, to improve performance and reliability of services and reduce unwanted side-effects, differential asset taxation is key.

Until recently, such tax policies have been fragmentary, tentative and ad hoc, hardly recognized as energy policy. Now, however, innovative electricity, including small-scale decentralized generation close to loads, offers more cogent reasons and more attractive opportunities to integrate and optimize entire local systems, including both generation and the technologies it drives. End-use technologies – lights, heaters, motors, freezers, electronics and so on – and even passive infrastructure such as buildings are part of the system that delivers comfort, illumination, motive power, refrigeration, information and all the manifold electricity services we take for granted. Upgrading end-use technologies is the most effective way to deliver better services more reliably at lower cost and with lower impact. But most electricity users don't know or care enough to do anything about it. Worst still, companies whose business is selling or delivering units of electricity want us to use more, not less. Inefficient lamps and motors benefit their cash flow. More appropriate tax treatment of electricity assets, especially end-use technologies and buildings, can provide a potent incentive to invest in upgrading infrastructure. It can give the incentive specifically to those whose business it is to deliver better services.

Users do not want reliable 'electricity'; they want reliable electricity services. With local generation, under local control, driving local technologies, the responsibility for keeping the lights on can be similarly local and coherent, and accordingly much more manageable. Moreover, this responsibility can be the focus of well-defined contracts between those enjoying illumination and comfort and those providing them, a more stable, less nerve-racking business than competing to sell ephemeral units of electricity.

This story feels much more coherent and convincing. Regard electricity not as a commodity but as a process delivering services. Improving the whole process benefits reliability and quality of service, while reducing vulnerability to disruption. It also offers the potential to shift progressively away from fuel-based to infrastructure generation, a key to sustainable electricity. If we can set this in train for electricity worldwide, we may eventually begin to recognize that all energy services, even including transport, are not commodities but processes. The challenge is always to optimize the entire process – an inherently positive undertaking for human society.

How do we get there from here? We have to start changing the way society thinks about electricity, and about energy. Everything else follows. That's why we have to get the story right. That's where those of us who are policy academics, analysts and commentators come in – right at the beginning, right here, right now.

10

GETTING ENERGY RIGHT

I don't like 'energy conservation'. I don't like 'energy efficiency', either. I've disliked them for more than thirty years. Come to that, as I indicated in Chapter 1, I don't even like 'energy' – at least not the way it's usually used. However, before you get the wrong idea, let me make myself clear. What I object to is the expression 'energy conservation'. I'm a physicist. Energy conservation is already the law, a law that can't be broken. I mean the first law of thermodynamics. Energy can never be created nor destroyed. In any physical event, energy is *always* conserved. Why, then, have we spent more than three decades advocating 'energy conservation', and lamenting that it isn't happening fast enough? We don't, of course, mean 'energy conservation'. We mean 'fuel conservation'. The distinction matters. Fuel is not energy. Energy is not fuel. If we smear them together as we so often do, we lose sight of differences that are crucially important.

This is just one of many ways we still get energy wrong. We use sloppy terminology that confuses and muddies the language of energy policy. That matters, not just to pedants like me: because it blurs and obscures the real options and opportunities we have. We need to understand the ways we have energy wrong, and what we might do to get it right.

Of course the first thing we have to do is to find people who *want* to get energy right – people with a professional interest in getting energy right, and the skills and knowledge to make it happen. But let's not delude ourselves. A lot of people, also skilled and knowledgeable, have absolutely no interest in getting energy right. They are already doing very nicely by getting energy wrong. They, their companies and their organizations have become powerful and influential, disproportionately so, precisely for this reason. They are more than content to let society go on getting energy wrong; indeed they throw up major obstacles to

This chapter is adapted from 'Getting energy right', London Boroughs Energy Group Christmas Meeting, Southwark Cathedral, 9 December 2005.

keep society from changing its approach. The obstructions and impediments they create are entirely understandable. If we start getting energy right, many people and companies will find their power and influence waning. But we must not allow them to shape policy for their own narrow benefit. Society demands and deserves better.

How, then, do we get energy wrong? To begin with, we describe the issue wrong. We persist in using language and concepts that drastically misrepresent what we are actually doing, so much so that we ourselves don't even understand what we want to do, or what choices we have about how best to do it. We even talk about 'consuming' energy. That expression makes my teeth ache. We don't – we *can't* – 'consume' energy. We 'consume' fuel. But we *use* energy. This all-too-common, wilful mistake typifies the central problem that underpins all the others. What we call 'energy policy' is nothing of the kind. It is 'fuel and electricity policy'. It is almost completely preoccupied with the supply of fuels and electricity, commercial energy carriers bought and sold as commodities in batch transactions, in which what matters is the unit price. This is an entirely valid activity, as far as it goes. But it ignores and takes for granted the entire reason for wanting fuel or electricity.

Modern fuels and electricity are useful only when combined with the necessary technology, the physical assets that actually deliver the services we want – comfort, illumination, cooked food, refrigeration, information and so on. If you already have the buildings, the lamps, the motors, the chillers, the electronics and other technologies, of course you want to be sure you can also get the requisite fuels and electricity to run them. But what you want is not just 'energy', otherwise unspecified. You cannot run your computer on natural gas, or your car on coal. Each particular end-use technology requires a particular form of fuel or electricity to run it, say high-octane unleaded petrol or 50-hertz 240-volt alternating current.

To smear all these different energy carriers together and call them all 'energy' suggests that they are all interchangeable. They are not – not without yet more technology, and sometimes not even then. Technology takes not only money for investment but also time, whether it is technology to produce an energy carrier or technology to use it. That time factor is important, because it usually implies a trade-off. If you want to ensure, say, comfort, in five years' time, you can use the intervening time in different ways. In many contexts you can either add more supply technology to deliver more energy carriers, or alternatively you can upgrade the end-use technology to make extra supply unnecessary. The same holds true for many aspects of energy use.

'Technology' includes all the physical artefacts that go to make up energy systems. The most important physical artefacts of all, the most important energy technologies of all, are buildings, for shelter and comfort and as platforms for so many other energy technologies. That is why I don't like 'energy efficiency'. You can measure fuel efficiency: it's the percentage of useful energy you get out of a device, compared to the amount of fuel energy you put in. But you can't measure the 'efficiency' of a building. A building intervenes in natural energy flows, and makes the temperature inside more stable and comfortable than the temperature outside. But you don't, indeed realistically you can't, measure the input energy, from sunlight, warm bodies, electronics and so on, or the 'useful' output energy either; so 'efficiency' is meaningless applied to buildings. Talk about 'energy performance', and avoid numbers that mean nothing.

Real energy policy must deal explicitly with buildings as well as all the other technologies – not just the technologies that produce and deliver energy carriers, but also the technologies that deliver the energy services we actually want. Real 'energy policy' is not just about commodity fuels and electricity. It is about infrastructure. Moreover, it is not just about so-called 'energy supply' infrastructure, such as refineries and power stations. It is about what we should explicitly and emphatically call the 'energy service' infrastructure.

Process and infrastructure

We see that most clearly when we consider electricity. At the beginning of the 1990s, when the UK government liberalized electricity, it imported the language and concepts of the hydrocarbon industry. It made electricity into a quasi-commodity, bought and sold like barrels of oil, in batch transactions in a competitive 'electricity market'. The objective of introducing competition was to make the price of a unit of electricity, a kilowatt-hour, as low as possible. It still is, as if this were what users wanted. It is not. What users want is a low electricity *bill* – not the same thing at all. What users also want is reliable services – keeping the lights on. We now know, as we might have anticipated, that a low unit price and reliable electricity may not be compatible.

No matter what free-market ideologues may say, electricity is not and never can be a commodity like oil or gas. Electricity as we use it is a process, taking place simultaneously throughout a complete system of physical assets – generation, network and loads. You can have electricity without fuel, but not without infrastructure. Electricity exists only in infrastructure. We must shake off the misguided belief that electricity policy can be treated as commodity policy, and electricity business as commodity business. Electricity policy should be first and foremost infrastructure policy – that is, policy to guide and shape the evolution of the physical assets that use electricity to deliver the services we want.

We already have an array of policy levers available to guide, shape and upgrade our energy service infrastructure. We should begin by recognizing explicitly that they are energy policy levers, more important than any policy levers that focus purely on fuel or electricity or their price per unit. We keep hearing that only 'higher unit prices' will make us improve our use of energy. That is simply nonsense. Minimum performance standards, planning requirements and government procurement are all potent policy levers to improve infrastructure. Perhaps most potent of all is asset taxation, the tax treatment of buildings and other energy technologies. But tax policy worldwide still stubbornly persists in a fundamental flaw, as we saw in Chapter 1. If you invest in an asset to sell its measured output, say a refinery or a power station, government considers it a business asset, and taxes it accordingly. If, however, you invest in an asset to deliver an energy service, say a well-insulated, comfortable home, it is not a business asset. You don't measure or sell by the unit its output of 'comfort'. In general, governments treat such investments much less generously for tax purposes than business investments whose outputs are to be sold. That one single assumption skews all investments in energy infrastructure, in favour of investments to supply saleable energy carriers and away from investments to deliver better energy services. Of course we do sometimes get grants, tax breaks, rapid write-offs, or other financial and fiscal benefits for improved end-use technologies. Unfortunately, however, such measures tend to be ad hoc, short term, inconsistent and incoherent. They have never been subsumed into coherent policy for energy infrastructure. They should be, as a matter of urgency, starting immediately.

Energy security means reliable services, not just fuel and electricity. The best way to minimize vulnerability to price rises or power cuts is to minimize dependence on fuel and electricity, by upgrading the infrastructure. The UK government, for example, has tens of thousands of buildings all over the country that are its own responsibility – everything from offices to prisons. Suppose it decided to upgrade all these buildings – to ensure they were solidly built and adequately insulated, with high-performance lighting, heating, ventilation and electronics, and indeed with on-site generation and cogeneration of their own heat and power. Such an undertaking would entail major pump-priming contracts to energy-service companies. It would boost skilled employment all over the country. It would cut the unit costs of the necessary technologies, by tooling up for expanded use. It would reduce the vulnerability and improve the reliability of all the energy services provided. It would demonstrate dramatically the vast potential for improvement. It would be a spectacular international public-relations coup. What's more, it would save taxpayers money. Imagine what such an approach could accomplish over the entire European Union, in OECD countries generally, or indeed anywhere in the world. Sustainable government procurement, not just of fuel and power but rather of energy services, could bring about a historic change of direction, towards real energy policy.

Real energy policy also points the way to real climate policy. You can put it like this:

- Climate is an energy issue.
- Energy is an infrastructure issue – not a commodity issue, an infrastructure issue.
- Therefore climate is an infrastructure issue.

To get climate right, we have to get energy right. To get energy right, we have to start by changing the way we think about it. We have to explain this change, with its profound and exciting implications, to the politicians, financiers and journalists that still fail to understand it. If we are serious about tackling energy security and climate change, if we are serious about keeping the lights on, we have to start by getting energy right.

11

SUSTAINABLE ELECTRICITY: CHANGING MINDS

Let me make you a proposition. Suppose I am an electrical engineering student. I have been assigned a project to design an electricity system. When I submit my design to my professors it proves to have the following specifications. It is based on large central-station generators, most of which operate either intermittently or at only partial load most of the time. The central-station generators that use fuel waste two-thirds of the fuel energy before it even leaves the power plant. The system necessitates long lines of network, in which line losses cost another significant fraction of the energy flowing. The configuration is inherently vulnerable to disruption, by mishap or malfeasance, over a wide area and almost instantaneously. It assumes that every load is essentially equivalent, requiring the same high quality of electricity. The system produces and delivers high-quality electricity as required by sensitive loads, much of which is then used for undemanding services such as heating and cooling. The generators are almost all thousands, more often millions of times larger than most of the loads on the system. Most of the loads are inherently intermittent or variable; but the system's large fuel-based generators are inherently inflexible.

Do you think my professors would give me a pass mark for my project?

Of course they would, if they were traditional professors of traditional electrical engineering. My project design looks exactly like the systems to which they have devoted their careers. Indeed it also looks like the systems to which three previous generations of electrical engineers have devoted their careers. That, needless to say, is our

This chapter is adapted from a presentation to Energy Networks Strategy Group Seminar on 'Networks for sustainable power systems – The need for change', Institution of Electrical Engineers, London, 2 February 2006.

problem. Our expectations are so low – not only those of electrical engineers but those of politicians, regulators, financiers, manufacturers and the general public. We put up with electricity system design and performance that would be embarrassing if we actually looked at it dispassionately, instead of taking it for granted. That is the overriding mental barrier we have to confront and overcome. We need to look again at what we want from electricity, and how best to get it, given what we now know about technology, fuels, finance, business, regulation and environment. One thing we know for certain: if we were starting from scratch to design the best possible electricity system with what we now know and what we now have available, my student project would bear little resemblance to the one just described.

'Best possible' here implies three criteria. The system would have to deliver what we want from electricity – call them electricity services – reliably, universally and sustainably. It would keep the lights on; it would keep everyone's lights on, everywhere; and it would do so indefinitely, without unacceptable side-effects of any kind, local or global. As yet we don't know in detail what 'sustainable electricity' might look like. But we can be pretty sure we know what 'unsustainable electricity' looks like. It looks like almost all the electricity we're now using, here and around the world. That is doubly frustrating, because we know we can do better – much, much better.

That, to be sure, raises another question. We are not starting from scratch – far from it, at least in OECD countries and most urban areas of the world. The mental barrier – the low and misleading expectations and all that goes with them – is accompanied by a major physical barrier: the existing system assets of generation, networks and loads, in their traditional configuration and their traditional role, the legacy assets that now look more like liabilities. They bring in train a legacy mindset that may be even more difficult to change. To tackle the mental barrier we must also work out a sequence of feasible changes, physical as well as financial and administrative, that will keep the lights on even as the system evolves to become more reliable, more universal and more sustainable. Unless we can imagine, in detail, how to get there from here, the mental barrier will remain insuperable.

Start with a vision

First we need a vision of what a sustainable electricity system might look like. Only if we have such a vision can we judge whether we are moving in the right direction, whether any given decision advances or retards our progress. Then we need a different way to think about what we are doing with electricity. Liberalizing electricity as we have done it around the world is based on a fundamental misconception. The electricity liberalizers tried to establish a 'market' treating electricity as though it were a fuel, a commodity. The result has been less than satisfactory, for a lot of reasons; but the underlying reason is basic. As we have emphasized repeatedly in earlier chapters, electricity is a process. It exists only in the infrastructure of assets that generate, deliver and use it, and through which it flows. Electricity is a function of infrastructure. Understanding this is the key to the necessary changes. You can produce and use electricity without fuel, but not without infrastructure.

The main mental barrier we have to overcome is to recognize, belatedly, that we have to stop treating electricity as a commodity issue. Electricity is an infrastructure issue. What matters most of all is the physical infrastructure within which we use electricity. We should think of it accordingly. Governments should make policies, and companies make plans, accordingly. That applies to generation, to loads, and – above all – to the network that links them. The infrastructure is the system – the whole infrastructure, including not only the explicit loads such as lights, motors and electronics, but also what we might call the 'implicit loads', especially the buildings. Once you start thinking this way, you rapidly find yourself in fascinating new territory.

For a vision of sustainable electricity we can start, paradoxically, with a building. You buy it as an investment. It will contain a variety of active technologies, and entail maintenance and running costs; but focus here on the physical asset that is the structure of the building itself. We know that the better the structure the lower the maintenance and running costs. The physical infrastructure of the building converts and reorganizes energy, particularly in the form of heat, as an inherent part of the function of the building. You neither measure nor pay for the flows of heat into, out of and through the fabric of the building. They are simply part of the performance of the building as built infrastructure, designed, paid for and used accordingly. The structure itself delivers, among other things, energy services such as comfort, as and when you use the building. The energy service is not a commodity. You do not measure it or pay for it by the unit. The physical asset, the building, delivers the service continuously, as a function of infrastructure.

You therefore design and construct the building to take maximum advantage of natural ambient energy flows, including light, heat, and convection currents of air. Daylight reaches much of its interior. The building structure acts as a heat store, soaking up excess heat or releasing stored heat, to keep the interior at a comfortable temperature whatever the temperature outside. Air circulates throughout the building gently and continuously, because of its interior layout and the small temperature differences between various parts of the building. These principles underlie a rapidly growing number of modern buildings in many parts of the world, and, indeed, many very old buildings, constructed before electricity became an option, when builders made skilled and subtle practical use of the light, heat and air circulation naturally available.

The building itself provides much of the illumination and comfort required. Electricity has much less to do to augment these services. The interior illumination comes from high-efficiency lighting. You use compact fluorescent lamps or light-emitting diodes, LEDs, for all lights that are on more than briefly, or are in awkward locations where you might otherwise have to replace burned-out incandescents frequently. Well-designed fittings direct the light to where it is wanted, with minimum waste. In temperate and tropical climates, buildings appropriately designed and constructed can maintain comfortable interior temperatures with little or no assistance from electricity. If your building needs active air-conditioning, it is badly designed. If, however, your building is in a cold climate, with outdoor winter temperatures well below freezing, you provide additional interior warmth by microcogeneration or by heat pumps, already common for instance in Scandinavia. You equip the building with sensors that measure temperature, light levels, occupancy and other relevant information. Automatic controls adjust the comfort and illumination to the levels you desire.

Inside the building, electric motors deliver motive power. You choose motors of the appropriate size, not oversized as is otherwise common, with variable-speed drives to maximize efficiency over most of their operating range. You also choose appropriate sizes of equipment they drive. You integrate fans and pumps, for instance, into ducts and pipes laid out to minimize frictional and other losses. You can therefore choose smaller fans and pumps, and smaller motors to drive them. Improving system design, integrating and optimizing the performance of all components together, gives you higher reliability and lower environmental impact, while costing less.

Electronics are designed to optimize performance and minimize losses. They handle information, including communications, data-processing and entertainment. Digital networks link together telephones, faxes, computers, television, video, building controls and security systems, not only within your own building but across cities, countries and continents. Note, however, that the stand-by mode for electronic equipment, such as the red light that stays glowing on the television when you use the remote control to turn it off, is a surprisingly heavy user of electricity in modern society. If you are aiming for high reliability and low impact, you redesign standby modes to reduce this insidious drain on electricity. In any case, when you are not using them, you just turn things off completely.

The building itself, the lighting, the motors and electronics all contribute to the electricity services you want. But they still need electricity to run them. Aiming for high reliability and low impact, you generate much of this electricity in or near the place it is used, minimizing vulnerability to disruption. Your building may have a gas engine, Stirling engine, microturbine or fuel cell in the basement, fuelled by natural gas, possibly operating in cogeneration mode, delivering electricity, heat, hot water or steam as you require. If your building is on an industrial site needing a lot of heat or steam, your cogeneration plant may include both gas and steam turbines. If your building is, say, a supermarket needing heavy-duty refrigeration, you can use not cogeneration but trigeneration, delivering not only electricity and heat but also ice-water for chillers. Trigeneration can raise overall fuel efficiency yet higher.

You can also equip your building to utilize not only ambient heat and light but also innovative electricity. As well as, say, the gas-fired fuel cell in the basement, in the skin of the building you incorporate photovoltaic (PV) tiles and cladding. You include three sets of cabling from the outset, one for telecoms, one for synchronized alternating current from the external network, and one for low-voltage direct current from your on-site fuel cell and photovoltaics. All computers and other electronics, the most sensitive loads in the building, require low-voltage DC. Traditional electronics need to include a heavy transformer 'power pack' to convert external AC to low-voltage DC, with accompanying heat, fan noise and losses. But if you operate them on your own low-voltage DC you can dispense with the power packs. Moreover, power electronics keeps the supply more stable and reliable than the electricity from the external network. Your high-performance lighting likewise operates on your own low-voltage DC. So long as the lights stay on, the computers and other equipment work, and the building remains convenient and comfortable, you may not even bother to measure the electricity flowing through your local DC system. You will measure and pay for the gas flowing into your fuel cell; but you may not measure the electricity coming out, nor that from your PV panels. You certainly will not pay for it by the unit. Instead, this unmeasured but functional electricity becomes analogous to the heat flowing through the fabric of the building. The electricity becomes part of the function of the infrastructure, part of the way the infrastructure delivers the energy services you desire.

This possibility brings in its train a startling corollary. Official prognoses of the future of world energy, from organizations such as the International Energy Agency, the World Energy Council, and the International Institute for Applied Systems Analysis, all indicate that whatever happens to world use of coal, oil and natural gas, world use of electricity will continue to increase essentially without limit, at least throughout this twenty-first century. They base their prognoses on so-called 'energy statistics', gathered assiduously by governments all over the world. Indeed, the word 'statistics' means 'information of interest to the state'. But consider your high-performance building. You don't need to measure the amount of electricity flowing through your local DC system. The government is certainly not going to measure it. Over time, if such local applications of infrastructure electricity, unmeasured and taken for granted, become commonplace, the statistical data available to record 'world electricity use' will become progressively as meaningless as any record of 'world heat use'. The implication underlines dramatically the distinctive difference of electricity, especially innovative electricity, as an aspect of energy policy.

Expanding the vision

We can now expand the vision of sustainable electricity. A wider network connects your high-performance building and other buildings on the system. It includes larger generators, notably large old hydro plants and modern microhydro units, as well as larger cogenerators, offshore wind plants and some generators in rural areas, such as village-scale biomass power plants. The wider network has some sections operating as synchronized AC. But it also includes many AC–DC–AC links that use power electronics to transfer electrical energy while blocking AC disturbances. The wider network is also heavily instrumented, and carries continuous real-time two-way information, not only about flows of electricity but also about flows of value through the network. The instrumentation keeps the network stable. It controls not only generators but also suitably flexible loads, also instrumented with embedded microprocessors and other controls. Instead of always increasing generation to follow load, the instrumentation may also reduce or disconnect flexible loads to follow generation, as appropriate. It also keeps track of network services, who is providing them and who is using them – who pays and is paid, and how much. Remote rural areas are not, however, connected to this network. They rely on self-contained local systems, that may include biomass power, wind power, photovoltaics and possibly batteries for storage, as well as appropriate system monitoring and control, also local.

As we envisage this system, aiming for high reliability and low impact, we know it is already technically feasible. It still has the wider network for backup and arbitrage, and to help ensure competitive pricing, for commodity or asset transactions. But it minimizes reliance on remote generation by large-scale hydro, coal-fired and nuclear power, and long-distance transmission. Accordingly, it is less vulnerable to disruption and reduces the environmental impact of electricity. Indeed it is not an 'electricity system' but a 'services system', with contracts and prices to match. But it is a long way from what most people still think of, when they think of electricity. That mental barrier remains the key problem.

The technical, institutional and financial arrangements of existing electricity systems reinforce each other. Any possible alternative starts at a deep disadvantage. We still assess even the anticipated costs of alternatives according to criteria used for existing systems, which may be completely inappropriate. How can you compare the cost of electricity from a remote coal-fired power station with the cost of electricity from a photovoltaic roof on the building where it is to be used, without mentioning location, time of day and year, accounting basis or other system costs and risks, including the risk of disruption? The comparison is grossly distorted. Nevertheless, traditional electricity people still use it to dismiss innovative generating technologies, by assuming they must fit into existing systems in the existing context. In any case, you must also ask 'The cost to whom? On what financial basis? Who carries the risks? Who pays for the environmental impacts?' Until we rectify such distortions, traditional assessment procedures will always undervalue and underestimate innovative electricity.

Invisibly sustainable

The system we envisage for high reliability and low impact embraces all the hardware that delivers the services, even including the buildings, not just the electrical hardware. Portraying such a system may take us part of the way towards a picture of sustainable electricity. But our portrayal omits salient aspects, especially the institutional. As yet, in such a system, we simply do not know how we would make decisions. We can anticipate only that we would no longer make many important decisions centrally, or at a national level. We would make many more essential decisions about electricity either locally or internationally than we have done hitherto.

Sustainable electricity may actually play a substantially smaller role in society. Services that electricity used to deliver no longer require it, or require much less. Moreover, as we noted earlier, official energy statistics may not even measure electricity we generate, say, by a photovoltaic roof, and use on the same premises, any more than they measure the daylight and body heat and other on-site energy sources that simply arrive and contribute their services. This kind of change is not merely an alteration of the structure of the electricity system. Sustainable electricity means a radical change not only in physical structures but in the way we think about electricity.

'Sustainable' implies not only environmentally and socially sustainable, but also economically and financially sustainable, and for everyone everywhere. If we construe sustainability as a counsel of perfection, we shall never achieve it. It can emerge only gradually, in electricity as in every other aspect. Minds, however, can change rapidly. If we stop taking traditional electricity for granted, if we begin to examine and question it, we may come to prefer innovative electricity. Changing the way we think about electricity will affect profoundly the decisions we take henceforth. Traditional electricity assumes centralized decisions. Innovative electricity does not. If many participants with many agendas make their own decisions, the effect may be untidy but dramatic.

If we get this right, if we overcome low expectations and a legacy mindset, our grandchildren may discover that sustainable electricity is invisible. No one measures electricity, buys it or even notices it. Infrastructure keeps the lights on.

Annex 1

RUNNING THE PLANET

Introduction: Seizing the millennium

The millennium is an arbitrary moment. It is determined by one cultural strand of many on the planet that all humanity shares. Nevertheless, precisely because so many people notice it, the millennium is an occasion to take stock of where we are, where we have come from and where we may wish to go. It is an opportunity to look beyond the immediate controversies and conflicts that preoccupy us, to take a wide overview with a long perspective.

We see the world swept by hectic change. But appearance may be deceptive. After tens of millennia, is human life significantly different, for individuals and society? Do we behave differently? Some would say not. But our behaviour may now have consequences more far-reaching and rapid than ever before. A single individual act can now produce an effect at once global and almost instantaneous. Moreover, people almost everywhere can learn of the act and its effect, as it happens. Human society now has technologies that magnify our behaviour, bringing us close together, whether we like it or not.

We depend, as we have always depended, on the life-support systems of earth, water and air, plants and animals, the living envelope of the planet. Human invention has also created other systems, physical, social, economic and political, on which society now relies. In some parts of the world, and for some people, these systems appear to work very well. However, in far too many places they do not, condemning far too many people to poverty, disease and brutal hardship. Moreover, evidence now suggests that both human systems and natural systems may be vulnerable to serious disruption. The global reach of our behaviour, and the pace of change of human systems, especially technological and economic, may be overstretching the capacity of our social and political systems to adapt. Natural systems themselves, taken for granted throughout human history, are showing signs of stress.

We humans now number more than six billion, and our numbers continue to increase. We also move around, more of us, farther and faster than ever before. We meet each other, not only indirectly but often face to face, across vast gulfs of difference in culture and circumstances. We speak many different

---

This annex, an essay for the new millennium, was originally published as a special supplement to *The World Today*, November 1999, Chatham House.

languages, explicitly and implicitly. All too often we do not understand each other. Understanding is difficult; ignorance and hostility are easy. We see the results wherever we look.

For more than three millennia people have been trying to develop and communicate a better understanding of the world and the way we live in it. Sages, priests, philosophers, historians, scholars, generals, artists, scientists, jurists, economists, all have created disciplines of thought and language to describe, discuss and direct human experience. These disciplines evoke great concepts – identity, love, family, nation, state, government, law, science, commerce, peace, war – and continuing debate. They resonate through our lives.

Nevertheless, as our numbers increase and even individual behaviour can have global consequences, the structures and processes of these human disciplines face ever-greater challenges. All too often they focus narrowly inward, when they should interact and complement each other. Dealing more and more in abstractions, aggregates and averages, they may lose sight of the individual. But each of us is an individual in society. Each of us acts as an individual, but is caught up in a nexus of interactions. In the formal world of states and companies we may be active or passive participants, but we participate. For good or ill, we cannot opt out. International affairs are also our affairs.

We have seen the Earth from space. We know how beautiful it is, and how finite. Against the background of the stars, we can recognize a kind of totality, here on Earth. As the arbitrary moment of the millennium catches our collective attention, the time is ripe to pool our efforts from every discipline. We need urgently to reassess the fundamentals of human life – from first principles.

The human problem

As the Voyager spacecraft left our solar system in the mid-1990s, it sent back a photograph of where we live. The Earth appeared as a blue dot. More than six billion of us now live on this blue dot, and it is getting crowded. We share the planet with other life forms, from blue whales to viruses; but we humans are now the dominant species – or at least we believe we are.

The dominance comes not from our numbers, but from our impact, as we change the face of the Earth to suit our purposes. We can influence how the planet itself works. From the inception of agriculture more than ten millennia ago, we have reshaped our surroundings and shifted their balance. We rearrange the soil and redirect the water. We foster the species we favour, such as rice, wheat and sheep, and diminish others, such as redwood, orchid and rhino. We alter the very air we breath, adding gases that affect our health and disturb the climate. The human capacity to create and destroy is reaching bewildering levels. One way and another we can now make parts of the planet sumptuous, for those fortunate enough to live there. We can also make parts of the planet uninhabitable, for ourselves if not for cockroaches.

We may be the dominant species, but some of us are more dominant than others. Partly for that reason we can and do make parts of the planet intolerable for each other. We have been doing so throughout human history. Why humans behave this way, and what to do about it, has been a central theme of thinkers at least from Homer onwards. Each individual has a personal agenda, starting with survival and continuing through satisfaction. How this works out in practice depends on the individual and the context. The thread from personal to general is always tangled; but it may now be long enough to circle the globe.

We interact with each other – sometimes in concert, sometimes in conflict. Humans are social beings. All of us need others, at least some of the time. We cooperate easily, at work and at leisure, in endless variety, from couples to multitudes. But cooperation can also dissolve readily into conflict, at any level – personal, tribal, national, international or global. Cooperation and conflict coexist. A degree of conflict – a difference of understanding, viewpoint or opinion – can be fertile and constructive, within appropriate limits. A conflict resolved to mutual satisfaction can enrich both sides. A conflict unresolved, however, can fester. Even at a personal level the result can be poisonous. Between groups of people the result can be devastating, both metaphorically and literally. All too often the ultimate outcome can be summed up succinctly: 'My survival or my satisfaction requires your death.'

For at least ten millennia this scenario has been taken for granted, as an everyday aspect of human experience. It has been played out in countless encounters, within and between families, tribes, nations and states, at every level of human society. It has been acclaimed and applauded, recalled and celebrated, the joyous triumph of us over them. It can now be played out globally. It no longer requires confrontation face to face. Destruction and death can be delivered with little if any personal involvement. A local conflict, even one initiated at a personal level, can now have global and almost immediate repercussions.

The problem of conflict, and the resolution of conflict, is our problem, and we are the problem. The overriding challenge facing humanity is to find a way to run the planet without destroying ourselves or the systems we depend on.

Systems for life

We are immersed in systems. Every individual human body is an unimaginably complex system that can feel, think, love and kill. It interacts with the systems of nature around it, maintaining temperatures and pressures, breathing, seeing, and sensing its surroundings. We manage this complexity without even noticing, unless the system malfunctions. If it does, if we become injured or ill, we turn to those who specialize in studying these particular systems. Over more than two millennia they have built up a parallel system of concepts and language to describe and analyse the system of the human body, and to identify and prescribe ways to forestall or rectify malfunctions – medicine, surgery and health care.

Those who study other natural systems, again with appropriate concepts and language – think of astronomy, geology, physics, chemistry, biology and all their many permutations – are not so preoccupied with malfunctions. Their first intention is to understand these systems and how they work. Some nevertheless seek ways to intervene, to alter the course of a natural system towards a different result, which may yield an outcome more desirable by some human criterion – a different material, a different process, an outcome specifically useful for some human purpose. Ever since humanity learned to control fire, the urge to intervene in natural systems has played a central role in human life, shaping not only our surroundings but the way we think and talk about them.

As it is with nature, so it is with human society. Any group of humans, even as few as a couple, constitutes a system – a network of interactions, a dynamic process that may or may not achieve a kind of equilibrium. Through the millennia we have evolved a multiplicity of concepts and languages to describe and analyse how humans interact in social systems. We can discern various social boundaries, the individuals and groups within them and the interactions across the boundaries. We can identify at least some of the forces that drive social systems – why people acting in groups behave as they do, in cooperation and in conflict. But we still do not know how to establish and guide processes that will keep social systems stable, within tolerable limits for the people in the systems, over the whole of the Earth, indefinitely.

We have already come a long way. After many tens of millennia of very gradual change, the organization and structure of human society have evolved at accelerating speed, especially throughout the century now closing. The first step of social organization is the family – humans directly linked by mating and birth. As the linkages become more extended, a family becomes a tribe, in which genetic relationships between individuals may be tenuous or non-existent but in which a shared history maintains the linkages. The word 'nation' continues the theme of extended family. Etymologically, a nation is a group of people linked 'natally' – that is, by birth. In practice, of course, the actual genetic relationships involved may be vanishingly small when not completely absent. At this level of complexity the linkages are not so much genetic as cultural – a shared history, language, modes of dress and behaviour. Moreover, at this level of complexity another aspect of social organization becomes obtrusive, if not indeed dominant. A family is by definition exclusive; you either belong to it or you do not. If you are not one of us, you are one of them, a simple fact of life. A tribe is exclusive more by choice than by genetics; but for that very reason the division between 'us' and 'them' becomes much more a factor in conscious thought and action. Carrying the process to its culmination, a nation defines itself as much by what and whom it excludes as by what and whom it includes. As a concept, a nation is an inherently divisive level of social organization, with potent consequences for the systems of human society.

The processes of social interaction within and between families, tribes and nations have always included both cooperation and conflict. Cooperation can be mutual, but even a cooperative relationship may be stratified, with clearly identifiable leaders and followers among those involved – even within a family, where patriarchy and matriarchy have long been acknowledged structures. Such stratification, in turn, may be established and maintained only through continuing conflict, between leaders and those who would replace them. Managing such conflict, and keeping it within acceptable bounds, has always been a key issue for social systems. Much of human history is a record of the process and its consequences.

The other most dominant theme of human history is the record of conflict between nations – that is, between groups of humans that identify themselves as nations. Such conflict is 'international' – between nations. Until less than three centuries ago nations in conflict were usually defined by their leaders – emperors and empresses, kings, queens, or other singular individuals sometimes accorded the status of divinity, as a kind of transcendental 'head of the family'. The gradual rise of the 'state', especially the popular state that emerged from the American and French Revolutions two centuries ago, changed not only the terminology of discourse but the concepts employed. A state is a significantly different social system from a nation. Its identity and coherence have different foundations, and its roles and functions are different.

A nation is self-identified by language and culture, a group of people that feel themselves to be a homogeneous community. A state, by contrast, is an institution to administer the society under its aegis. Some commentators argue that the role of the state is legitimized by the fact that it administers a nation. A government may then allude to 'the nation' it administers, to underpin its legitimacy. Conversely, a nation can argue that it is entitled to a state. But the nuances implicit in the concepts of 'nation' and 'state' remain both potent and confusing. For historical reasons, for instance, the Charter of the United Nations, like the name of the organization, refers to 'nations', not 'states'; but the members of the General Assembly and the Security Council are states.

The term 'nation-state' attempts to elide this difference; it also avoids the ambiguity arising in federal structures such as the United States, in which a 'state' means a regional government subordinate to the federal or 'national' government. Nevertheless, to refer to a 'nation-state' may seriously blur distinctions that ought to be clearly recognized, that are growing steadily more important – not least in the area of conflict, and resolution of conflict. Nations and states overlap, but rarely now coincide. Not all nations have states; few states now administer distinct nations, in the original sense of the concept. With some reservations, this discussion will use the word 'state' as it applies to members of the UN General Assembly.

The systems of human society – the social systems, the political and economic systems, and the technological systems – now cover the entire Earth. However, even where they usually work well enough they are now prone to breakdowns, and threatened by ever-graver malfunctions. These system problems are not just theoretical but practical; and they need solutions – coherent, global and lasting. The need is urgent, and time is short.

Global society

Human society itself is already global. Its structure and organization embrace the entire planet. Deep in a remote rainforest, activists with satellite cellphones track illicit logging, in real time and in contact with colleagues thousands of kilometres away. The geographical separation and the traditional social stratification of human activities and interactions are blurring and fading, giving way to a complex dynamic network whose fluid linkages no longer fit easily into conventional formal frameworks.

Formally, nevertheless, the six billion people of the world are still organized into a framework of states. According to one definition widely accepted, each state has a defined territory within identified borders, a name, a population of citizens and a form of government recognized by other states. Perhaps 200 such entities are now variously recognized worldwide. The government of a state, a 'national government', however constituted, is considered 'sovereign', with certain privileges both inside and beyond its borders. However, the practical meaning of sovreignty, as it affects the behaviour of individuals and groups representing a national government, has always been a matter of dispute.

A national government, for example, exercises various powers over the individuals within its borders. The ultimate manifestation of these powers is that the government can claim to hold the monopoly of the legitimate use of violence against those within its borders, including the use of police, courts, prison and lethal force. As recent events in Kosovo, Rwanda and elsewhere have demonstrated, this claim, like other attributes hitherto accorded to governments of states, may require reassessment; but the process will be fraught with pitfalls.

Within a state, other levels of government subordinate to national government may exercise significant responsibilities, including regional government of a province, subnational state or land and the local government of a city or municipality. The allocation of powers and responsibilities between different levels of government within a state is usually contentious and continuously evolving. The United States of America, for instance, as its name indicates, was created by a merger of states previously governed separately under the British. Tension between the US national or federal government and the governments of the separate states continues more than two centuries later. Similar tensions arise in other states with formally federal systems, such as Australia, Brazil, Canada, Germany and Switzerland. Even city and municipal governments can confront their national governments on issues of power and responsibility.

More recently, states themselves have begun to surrender a measure of their sovereignty to supranational alliances with certain powers over even national governments. The most fully established is the European Union, with fifteen member states and a queue of others waiting to join. MERCOSUR in South America and the North American Free Trade Agreement NAFTA are less fully fledged but developing. The World Trade Organization may be the first worldwide supranational organization with powers to compel all member states to comply with its rulings, and to impose sanctions for failure to comply. Although the remit of the WTO is global, its actual membership is still patchy. Nevertheless, despite the surrender of sovereignty entailed, non-member states including China appear eager to join. States themselves are now thus interacting in ways that weaken once sacrosanct national borders. Other dimensions of social organization are reinforcing this trend, especially those groups of individuals cooperating in the form of companies and corporations.

States still presume the power to defend their borders, to tax the citizens within those borders, and to make and implement cross-border agreements with other states. However, as international trade and commerce expand and global telecommunications penetrate to the most remote corners of the Earth, many national borders are becoming increasingly permeable if not effectively imperceptible to many human activities. Information now flows freely around the world, even into and out of locations where governments may wish to impede its flow. So does money. Goods and people are not so mobile, but they too are on the move, often across borders impenetrable not so long ago. The gradual opening of national borders to people and their organized activities, including companies, complicates traditional government responsibilities such as taxation, management of national currencies, maintenance of national standards for goods and services, and control of crime. The fading of national borders weakens the state, the central framework of social organization upon which global society – or at least its historically most powerful elite – has learned to rely.

Until recent decades 'international affairs' always clearly involved two or more states, and could be readily distinguished from 'domestic' or purely national affairs. Now, however, as a result of international trade and commerce, international currency dealings and transnational corporate ramifications, almost any human activity beyond the level of subsistence farming has international connections, implicit if not explicit. As an aspect of worldwide social organization, government at whatever level may therefore be growing less important in practical terms, while companies and corporations, acting across national borders and indeed sometimes globally, grow more important.

A company or corporation is a 'legal individual', an entity recognized in law, made up of a group of people working together towards a shared objective, within an agreed legal framework of shared responsibility. Their individual responsibility, however, for instance to repay debts, may be limited by the law under which the company is recognized, a major benefit of incorporation for the individuals involved. Government of course makes the law according to which society recognizes a company, and which provides a framework within which the company acts. But the activities of many companies can be and now often are much more far-reaching than those of most national governments – geographically, financially, and in impact on the daily lives of people.

Companies and corporations now represent a strong but quite different framework of social organization from governments – not necessarily limited by geography, without borders as conventionally understood, and now often extending worldwide, as multinational or transnational entities. Moreover, the nature and function of these entities are also changing. Many of their most significant economic activities now involve international transactions based entirely on flows of information, intellectual property, entitlements and finances rather than physical goods. Some commentators foresee the emergence of the 'virtual company', a global entity defined by networks of contracts and information – a form of social organization far removed from those based on genetics or geography.

Companies and governments interact, but have very different objectives, powers, responsibilities and scope. Almost all people are citizens, willing or otherwise, of one or another state and its national government. Most people on Earth, however, do not belong to any company or corporation – not even a local one, much less a global one. That is not necessarily a disadvantage; but it may yet prove to be, as patterns of power shift. Already it dramatically separates most 'haves' from most 'have-nots'.

As aspects of social organization, states tend to encompass millions of individual people, and companies up to many tens of thousands. For each individual, however, perceptible social organization begins with identity – 'I' compared to 'you' – and extends, progressively, through family, friends, colleagues and acquaintances out to an anonymous world beyond, the wider social framework within which each person lives and functions.

Identity seen from inside looks very different to identity seen from outside. From outside, your identity will be described and labelled for social purposes: name, personal attributes – female, male, old, young and so on – and according to where you fit into the larger social framework: marital status, job if any, company if any, nationality, national origins, connections and allegiances with one social grouping or another. Although you may be able to choose or change your employer or the political party or football team you support, you cannot choose or change certain aspects of your social identity, for instance your national origin. You are identified as a member of a particular group, a subset of society distinct from others.

On one level, such social identification is a straightforward process of differentiation, to tell one person from another. On another level, however, it may implicitly if not explicitly carry value judgements about the comparative worth of individuals. Social identification is an essential tool of social interaction; but it may be an edged tool. It helps to determine not only what we do for each other, but what we do to each other.

Living together

Humans need each other; but individuals have different agendas. Over the millennia we have evolved an assortment of ways to resolve disagreements on how to behave together, how to reconcile different agendas, to maximize useful cooperation and minimize corrosive conflict. We have always had to manage this individually at a personal level. We now also have to manage this collectively, not only at local and regional levels but at a global level.

The crucial step is to acknowledge that no individual can behave without reference to others. Individual actions happen in a social context, and are shaped by that context. How you talk, how you carry yourself, what you wear, what you eat and drink, how you interact with others of your own or the opposite sex, or older or younger, or friends or strangers, all are part of the social structure and process, the culture in which you live. Much of this culture is absorbed in infancy and childhood, with little or no conscious thought. You take it for granted as the framework within which you behave, and you behave accordingly.

The social and cultural framework influences the decisions taken, and the consequent behaviour of individuals and groups. Some decisions are yours to take; some you can affect but not determine; some are taken for you. If you do not like a decision, you can accept it nonetheless, or protest against it; the nature of your response is an indication of your power. Your capacity to take, implement and enforce decisions leading to action is a measure of the power you command within your social structure and its processes. The stratification of a society is above all a stratification of power, of individuals and groups within the society.

From the beginning of recorded history the concept of power – why and on what basis some individual or group can compel other humans to behave a certain way – has fascinated commentators. One human can influence another's actions by direct physical threats, by rewards and punishments, or by persuasion, overt or subtle. Four categories of power able to influence both individuals and groups can be readily identified. The first is physical power – personal violence and lethal force, including military force. The second is productive power – control of resources and labour. The third is financial power – access to money and ability to grant credit. The fourth is intellectual power – knowledge and how to acquire, control and use it. In today's world, some would add a fifth category – cultural power, the ability to disseminate a worldview and affect the climate of thought.

Across the world, different individuals and groups have different levels and distributions of these categories of power. Away from the immediately personal, the most obvious locus of power is government. At the moment, people are organized into states with national governments, some with regional and local governments subordinate to them. The national governments currently recognized differ widely in the power they can exercise, both within and beyond their borders, according to the categories listed. Today much the most powerful state, taking all the categories together, is the United States. Where power will reside in the new millennium, however, is much less certain. The balance may shift not only between states – consider China, India and Brazil, to name but three – but also between states and global companies or other entities, with concomitant effects on human life everywhere.

The concept of 'government' itself already covers a wide range of roles and functions for those individuals who represent and exercise powers and responsibilities according to this concept, and for those upon whom these powers are exercised. Across a broad spectrum of human activities a government can tell people what to do and what not to do, can reward and punish individuals and groups, and in the ultimate sanction kill them. Profound questions arise afresh. On what basis can a government legitimately exercise these powers? Is this basis adequately valid now, and will it remain so? What constitutes 'good' or 'bad', 'strong' or 'weak' government?

Authoritarian government rules ultimately by force or the threat of force; changing an authoritarian government may involve violence. The alternative may be democratic government, 'of the people, by the people and for the people' or with the 'informed consent of the governed', in which a government surrenders power and leaves office without protest when it can longer command popular consent. But the term 'democratic' has long been routinely misapplied and misused. Moreover, even where governments change peacefully, and people give their consent through ballot boxes, the process of legitimation of government may fail to address fundamental practicalities of social organization, including access to finances and information.

These practicalities are growing steadily more demanding, even within states with well-established government systems and comparatively stable social order. Indeed, precisely where these conditions are most evident, people are turning their attention away from the processes of government. Many no longer even bother to vote. This should not, however, be construed as a sign of popular contentment. It may rather be a sign of disillusion with government and the political process that validates it, that gives it legitimacy. If so, the portent may be ominous. The discontented, too, now have access to power that can be significant, disruptive and difficult to counter.

Government, as it refers to the social process by which a certain group of people make decisions on behalf of a larger group, is a subset of the broader concept of 'governance'. The question of 'good governance' comes into play not only for entities of government but also for other social entities such as companies. What constitutes good governance, and how is it achieved? These questions are ever more pressing. For a company, those making decisions, the senior management, are granted the power to do so by those who own the company – perhaps shareholders or private investors. Once the senior management is in office, the owners withdraw to a passive role, awaiting the results of the decisions they have empowered the management to take. If they do not like the results they remove the existing management from power, and assign the power to others.

Much has been made of the putative parallel between a democratic government and a company owned by shareholders: the voters empower the government to manage the affairs of the state, and remove the government if it does not deliver acceptable results. This makes the voters by analogy the owners of the government. In any context, however, the concept of 'ownership' carries with it a cargo of corollary assumptions about power and its practical use, that must be examined carefully. Moreover, however powerful a company is, its very existence, recognition and continuation as a social organization depend on the government that creates and enforces company law. In any form of social organization, in any group of people acting together for whatever reason, from a family outwards, governance – who takes and implements decisions on behalf of those within the organization, on what basis and with what legitimacy – is a primary concern.

An authoritarian government or a dictator can exercise power capriciously, on a whim. People subject to this whim have no reliable framework of reference within which to behave. To reduce the consequent uncertainty and create a form of predictable order, society through its government creates laws. In principle these laws stand independent of the particular individuals or groups administering them. The body of law established in a society becomes a framework defining decisions and actions acceptable to that society, and penalties for decisions and actions that society considers unacceptable. Social criteria still differ widely. Many societies, for instance, have now abolished the death penalty for even the most serious transgressions; others retain it, some for offences that other societies consider minor. Nevertheless the rule of law remains a crucial underpinning of social organization, which otherwise deteriorates into an endless conflict of ruthless physical power, as too many recent examples demonstrate.

At the moment, the global mosaic of states also constitutes a global mosaic of laws. Any given law is ultimately defined and enforceable only within the borders of that particular state. So-called 'international law' is usually referred to as 'soft law'. It embodies agreements, conventions and treaties between national governments, which provide a quasi-legal framework for international activities. But it ultimately has no formal enforcement mechanism to compel compliance, nor any consistently effective international regime of penalties or other sanctions for non-compliance. Whether a particular international law can be enforced in a particular context depends on the power of the national governments involved, not on any international or global rule of law. Nevertheless the UN Tribunals on Rwanda and Bosnia, the WTO tribunal and the International Criminal Court may represent the first tentative steps towards a different regime. Extending the rule of law to encompass the planet as a whole will be daunting. In the new millennium, however, it may be essential.

Disparities and dangers

Every human individual is different and distinct. The fundamental difference every human notices is the difference between 'me' and 'everyone else'. That is the starting point for human relations, social organization, cooperation and conflict. Humans are endowed with memory. The accretion of remembered experience creates a sense of identity, at first personal, then social – shared identity, as one of a group. In today's world, almost every human individual has many group identities, overlapping but distinct. Subsistence hunters and farmers may identify themselves primarily within their families or settlements; but a well-to-do European or North American with a family, a job, leisure pursuits, an internet connection and a chance to travel may have dozens of different identities, depending on the context at any given time.

The sense of identity as one of a group establishes bonds between those in the group. But it also underlines differences, not only between 'me' and 'everyone else', but also between 'us' – the members who share the group identity, however recognized – and 'them' – everyone outside the group. Differences between individuals and between groups can be considered under two headings: diversity and disparity. Diversity, a portfolio of variously different attributes brought together, is beneficial; one attribute complements another, making the whole stronger than the parts. Disparity between individuals or between groups, however, implies inequality of some kind.

A group of people, for instance, with diverse experience and skills, combining this variety of experience and skills to act together, can accomplish achievements beyond the capacity of a group without such diversity. This group, however, may also manifest disparity. The activity to be undertaken may, for instance, be decided by only one or a few members of the group, without consulting the others; the experience and skills of the others are required, but not their views on the choice of activity. If those who decide on the activity can nevertheless induce or compel the others to help carry it out, the decision-makers are more powerful, in at least this context, than those who have no say in the matter. Disparity of power is an omnipresent attribute of human social organization, from families to states and companies. It is also a corollary of the many other disparities that permeate human society.

The concepts of diversity and disparity deserve careful examination. When is a difference diversity, and when disparity? Categories of difference between individuals include, for example, age, gender, appearance, health, strength, intelligence and competence, including social skills. Categories of difference between groups include number of members; language; shared history; culture; acceptable modes of behaviour within the group; acceptable modes of behaviour to those outside the group; access to resources; and collective competence, including the ability to act together coherently within the group. Such differences are easy to recognize, less easy to classify as diversity or disparity. A difference of gender, for instance, is clearly diversity, not least in the sense of 'vive la difference'; but it is also disparity, when opportunities unrelated to the biological difference of the genders are available to one gender but not to the other.

If difference is manifest simply as diversity, it enriches the collective human experience: 'variety is the spice of life', 'opposites attract' and so on. Difference, however, is often genuine and objectively verifiable disparity, for instance between the competence of a professional golfer and that of a duffer, or between the spending power of a Wall Street banker and that of a Russian peasant. Even when difference cannot objectively be called disparity, it may nevertheless be thought to be. Disparity may be simply perceived, or inferred as a value judgement: 'We are better than they are, because they don't speak the right language/wear the right clothes/say the right prayers...' Whether a particular difference is considered diversity or disparity, by whom and on what basis, has profound implications for social processes. So does the question of whether a disparity is genuine or merely perceived.

Disparities, whether genuine or perceived, shape interactions between humans and between groups. They affect individual self-interest and self-esteem, families, tribes, nations, states, corporations and other organizations to which humans affiliate or belong. At a global level, disparities are manifest not only within states but between states. For the stability or otherwise of social organization, disparities between individuals and disparities between groups are crucial, especially disparities of competence and of power, however manifest. Disparities all too often engender resentment and hostility, hatred and fear.

The planet has always been riven with disparities. Many we simply take for granted as a fact of life – disparities, for instance, of individual physical size and strength, appearance, applied intelligence or competence, and so on. Even such disparities can nevertheless cause distress and social tension for those on the wrong side of the divide. If you find yourself at a disadvantage because of some disparity, you may strive to gain a compensating advantage according to some other criterion. If you succeed, the first disparity becomes less important, and matters less to you. A great deal of normal social interaction consists of such implicit jockeying for position within a group. From outside the group, a pattern of such compensated and complementary disparities may simply appear as diversity.

A larger aggregation of people will include many such groups, some overlapping, many with little or no direct contact with each other. Individuals interact among themselves within the group; and each group interacts with other groups. Once again, the differences between groups can be considered as diversities or disparities. If differences are considered only as diversities, opportunities arise for cooperative and mutually advantageous interaction between groups. If the differences appear as disparities – if one group is, or considers itself to be, at an advantage or disadvantage relative to another group according to some significant measure – cooperation becomes more constrained, and tensions increase. The effect can be seen throughout society, within and between families, neighbours, neighbourhoods, local social groups of every kind, schools, clubs, companies, towns, cities, regions and – quintessentially – nations and states.

When considering the tensions in human society, however, a crucial distinction must be made between two categories of disparity: the objectively verifiable and the perceived. Perhaps the most extreme example of objectively verifiable disparity is that of access to resources. Some fortunate humans can pick and choose among an abundance of resources competing for their attention. Others, while not so fortunate, have adequate access, enough to satisfy what have been called basic human needs – shelter, clothing, clean water and food at a minimum, possibly extending also to health care and education and, by some criteria, employment. Many humans, however – certainly more than one billion, possibly more than two billion (a third of the people on Earth) – do not have adequate access. Their shelter and clothing are meagre, their water filthy, their food sparse and unhealthy, their lives unremittingly arduous and their numbers increasing. The disparity between their circumstances and those of the fortunate is gaping, and widening. The disparity between their power and that of the fortunate is also, at least superficially, gaping. But the poor are not entirely powerless. Their numbers alone mean they cannot be ignored. The coexistence of opulent luxury and desperate poverty, sometimes within the same urban area, on a finite and interconnected planet is not a recipe for stability.

The present-day disparity between rich and poor is a consequence of human social behaviour throughout past millennia. From one viewpoint human history is a record of people grouping together, cooperating to achieve common purposes. It is also, however, a record of divisive behaviour, of conflicts between groups set on widening the disparities between 'us' and 'them' in favour of 'us'. Historically the process has been a central organizing principle of society. It has entailed aggression by one group against another, invasion, slaughter, occupation of land, suppression of opposition, torture, colonization, exploitation and slavery, opening wide gaps between the winners and the losers. It has defined key factors such as ownership of property and materials, land tenure, water rights, and transfers of goods and services from the poor and powerless to the rich and powerful. The organization and structure of global society today is in large measure the legacy of divisive human behaviour through the ages.

A related organizing principle, dramatically effective especially for the past two centuries, has been the evolution from subsistence food production and craft manufacture, essentially self-sufficient but limited, to a division of labour. Division of labour is nominally a cooperative activity; diverse skills and competences are deployed appropriately to multiply the productive output of goods. These goods can then be made available to all participants by means of an agreed mechanism of exchange in the form of money, and a market-place in which exchanges can be transacted. In practice, however, division of labour usually leads to wide disparities between the participating individuals and groups – in particular, disparities of power and of access to resources. Why this is so has long been a topic of discussion and dispute. It became the focus of one of the fiercest global confrontations of the twentieth century, between the forms of social organization known as capitalism and communism. That confrontation has now abated; but the underlying question festers unresolved.

Some commentators have long advocated straightforward redistribution – taking from the rich to give to the poor, and thus in some measure making restitution for millennia of flow in the opposite direction. Although even some of the comparatively rich endorse this view, the practicalities are awkward, not least precisely because the rich are at once more powerful and usually reluctant to surrender either riches or power. Instead, in recent decades, the touchstone of social evolution on the planet has been so-called 'economic development'. Commentators argue that economic development will better the lot of the poor, and help to reduce the disparities between poor and rich. The proclaimed objective is not to impoverish the rich, but to enrich the poor. To do so requires, according to the analysis, creation of more 'wealth', making the global pie bigger, to allow a larger quantity of wealth to reach the poor.

Wealth, however, is yet another concept more easily employed than defined. It is a standard basis for comparison, and is routinely used to quantify disparities between individuals and groups. But its meaning has been gradually but inexorably evolving, and the process may be accelerating. In a simple society, wealth can be measured, for example, by the area of land or the number of livestock an individual or a group owns, however ownership is established and retained. As a society becomes more complex, the measure of wealth becomes more abstract. It is often summarized not directly in terms of ownership but rather by possession of a certain quantity of the medium of exchange – so much money, and therefore by implication the goods or services that money can buy. The transactions do not have to take place, and perhaps cannot, at least not readily, if the wealth in question is, say, largely in the form of the owner's house. Wealth itself gradually becomes hypothetical and inferred rather than demonstrable.

Layer upon layer of abstraction can be superimposed. The global transactions now taking place continually, in which one currency is simply exchanged for another, dwarf the global transactions in actual goods and services. These currency transactions, abstract though they be, alter the measure of value of goods and services. They thus profoundly affect the quantified wealth of individuals and groups, up to and including states. Measured in this way, wealth can vary substantially, even day to day and indeed minute to minute. What remains true nevertheless is that even the abstract quantified measures of wealth still exhibit vast and persistent disparities, not only between individual humans but between postulated 'average' citizens in different regions and states of the world.

Philosophers and other commentators throughout history have stressed that wealth does not necessarily make individuals happy or satisfied. The collective wealth of a group, from a family to a state, says nothing about the balance of wealth and satisfaction of its individual members, nor about the stability and durability of the group as a social organization. That said, the disparity between 'haves' and 'have-nots' grows steadily more acute. In today's interconnected world, 'have-nots' are made perpetually aware of their disadvantage, and 'haves' can no longer easily insulate themselves against nor ignore those less fortunate. Greed and resentment stirred intimately together make an explosive mixture.

As if genuine disparities were not problem enough, human society is riddled with perceived disparities – the 'us-them' syndrome. Perceived disparity, according to which one individual or group considers another inferior, can be based on almost any criterion, or none. Although it entails value judgements that may have no basis in objectively verifiable fact, it has been a feature of human life and human society at least throughout recorded history; it probably dates back to the emergence of the species. At its ugliest it leads to atrocity, 'ethnic cleansing' and genocide. If perceived disparity proves to be an inherent and uncontrollable part of human make-up, global society is in deep trouble.

Conflict does not require disparity; but disparity breeds conflict. If we are to manage and mitigate conflict we shall have to manage and mitigate the worst genuine disparity, wherever and however it appears. We shall also have to learn to recognize, to welcome and indeed to celebrate diversity, without perceiving it and stigmatizing it as disparity.

Knowledge, faith and wisdom

What do we know, and how do we know it? As we grapple with the challenge of running the planet, the process of acquiring and applying knowledge, individual and collective, is central to the task. The world is now awash in information. It is gathered assiduously and disseminated continuously, to almost every corner of the globe. But information is not knowledge. To become knowledge, information must be assimilated and understood, by individual humans and by groups of humans. To be useful, information thus assimilated and understood must be correlated with other knowledge. Only then can it influence human decisions and human behaviour. The flood of information now engulfing the world is meaningless unless and until it becomes knowledge. Nevertheless knowledge too – at least some categories of knowledge, in some places – is expanding faster than ever before; indeed it is becoming a key dimension of economic activity. But knowledge, both individual and collective, is ultimately and inevitably limited. Moreover, what we need most is not merely knowledge, but wisdom. Wisdom is all too rare.

A small child is usually a voracious and insatiable learner, acquiring language and behavioural competence at a rate that would be breathtaking in an adult. The learning process is both personal and social. Much of what a child learns is absorbed without conscious thought, including a personal identity, group identities and social skills – how to interact with other humans.

As yet, however, we do not understand the process of childhood learning well enough. Perhaps if we understood it better the older humans involved might not so often make a mess of it.

What we do know is that childhood learning is formative, exercising a profound and lifelong influence on the individual. What you learn as a child affects everything you do for the rest of your life, including the way you behave, the decisions you make and how you interact with other humans. Historically, the process of childhood learning has usually been deeply and inherently divisive. It stresses differences and exclusive identities – family, tribe, nation, the 'us–them' dichotomy. At best it undervalues the wider context, the system context, in which each individual interacts continually if mostly unconsciously not only with human systems that now span the globe but also with the natural systems essential to human life, the web of interdependence.

The process of childhood learning tends to see differences not as diversity but as disparity. The starting point is the disparity of size and power between the child and the adults it first encounters, as it establishes its personal identity; the child then seeks forms of countervailing power – disparities in its favour, including not only genuine but perceived disparities. Throughout human evolution this process has clearly had survival value, as different groups compete for resources and power and their progeny join the fray. In the coming millennium, however, the consequences may be crippling, not only for humanity but for the planet.

One process of acquiring knowledge is called education. Every human individual undergoes some form of it, which may or may not involve schooling, teachers and formal qualifications. Some societies routinely acclaim it as universally desirable if not essential. Others restrict access to some forms of the process, excluding those for whom such education is judged unsuitable, for example young females. Etymologically, 'education' means 'leading out'. In practice, it usually entails the transfer – or the attempted transfer – of competence and understanding from those more experienced and therefore commonly older to the less experienced and younger. The communication required may be direct and personal, or indirect through a medium that can convey a record of experience, real or imagined.

However education occurs, it conveys, at least implicitly, not only factual information but a particular way to organize and evaluate that information and correlate it with knowledge already acquired. The process of education is inevitably selective, not only because of the limited capacity of human individuals to assimilate knowledge, but because those in authority over the process may consider some knowledge undesirable, for example knowledge about human sexual behaviour.

Education, however arranged and guided, is but one aspect of the process of acquiring knowledge. Most knowledge is acquired more or less continuously if incoherently, as a result of everyday life and social interaction; much is trivial, some useful, some unquestionably influential on the individual or group acquiring it, prompting action or behaviour that otherwise would not occur. What any given individual knows is different and distinct from what anyone else knows; but groups of people share bodies of common knowledge. They compare, share and expand this common knowledge by communicating through explicit and implicit language. They can do so in two ways, whose approach, consequences and implications are essentially different. One can be called, broadly, the scientific method. The other can be called faith.

At its most basic, the scientific method is organized curiosity, prompted by an awareness of ignorance about something of interest. You don't know what will happen when you do something; so you try it and see. You believe the evidence of your own eyes and other senses. Then you do it again, and see if the result is more or less the same. If it is, you have acquired new knowledge. Through the millennia, humans have learned to control fire, to manipulate materials, to foster desirable plants and animals, to create artificial structures – a cumulative body of practical knowledge, skills and competence that now allow us to send human artefacts to the outermost reaches of the solar system and beyond. This practical knowledge has created much of the physical framework of global society today.

For much of human history this body of practical knowledge was largely accumulated and transmitted personally, for instance by artisans to their apprentices, as diverse forms of understanding without any unifying coherence to link one discipline with another. Only with the rise of so-called 'natural philosophy' – what we now call 'science' – did such unifying coherence begin to emerge. Natural philosophers and scientists created concepts, language and procedures that proved to be a powerful way to approach, organize and interpret human experience. One of the most striking consequences was to recognize the interlinkages and interdependence between apparently diverse and disparate forms of experience. As understanding extended to encompass aspects of nature beyond the reach of our personal senses, such as atoms and molecules, crystals and cells, disciplines once entirely distinct – for instance metallurgy and botany – revealed underlying commonality.

The body of knowledge now called scientific is already vast, and expanding faster than ever. To be accepted as scientific, knowledge must be objectively verifiable on some basis commonly agreed. Science as an approach to knowledge is both cumulative and collective. It is inherently social; it functions by communicating and comparing recorded experience, results and analysis, usually including measurement and quantification for precision where appropriate. Crucially, scientific knowledge is always and inherently provisional. If the accepted scientific knowledge proves inadequate to reflect and represent actual experience, the knowledge must be amended accordingly. If the evidence changes, the scientist's understanding has to change.

The scientific approach to knowledge has given humans significant capability to intervene in natural processes and redirect them to fulfil human purposes. For a century or more, the scientific approach to knowledge has also been applied directly to humans themselves, not only to the functions of an individual human body, with considerable success, but also to the functions and behaviour of humans in groups, not perhaps so successfully. Biochemistry, behavioural science and other disciplines have yielded fascinating insights, and suggested new ways to think, for instance, about illness, consciousness and intelligence.

As yet, however, these studies and findings remain intensely controversial, at best provisional even by the usual cautious scientific criteria. Disputes continue as to whether, for example, intelligence as manifest by humans can be meaningfully attributed to non-human animals, or to human artefacts such as computers. Science still has difficulty dealing adequately with living systems – that is, not just individual organisms, however identified and isolated, but entire systems of organisms, interacting continuously with their surroundings and each other in processes of extreme complexity and subtlety.

Some forms of science approach this complexity by arbitrarily reducing it to fewer components, breaking up the complexity into manageable subsets. Unfortunately, by doing so – for instance by killing the frog in order to dissect it – science loses an essential part of the relevant information. In many circumstances the most important part of the information is precisely this system dimension, the interaction and interdependence that is often literally vital. This has always been true of the human relationship with the natural systems that sustain us. It is now also true of the human relationship with human systems, including technology and social organization. Thus far, as an approach to knowledge, science has been much more effectively applied to natural and technological systems than to social organization. Disciplines such as economics, sociology and anthropology endeavour to study aspects of the behaviour of humans in groups, combining historical and observational evidence and scientific analytic techniques; but success remains limited.

Every single experience is unique, because of the individual who has the experience, and the effect of time. Everything that happens happens only once. From the scientific viewpoint this poses very deep problems, not least about consciousness, the sense of self and the sense of time. Nevertheless, biochemistry is now suggesting intriguing interpretations of psychology, human moods and behaviour, and even of intelligence and understanding, how they are manifest and applied. To some extent, many people now consider science to be developing an understanding of how the planet works, and even how the universe works.

Here, however, the crucial word is 'how'. For humans, understanding how is not enough. Humans want to know why. Why are we here? Why is life? Why is death? The greatest conundrum for human understanding is the fact of the end of life, including your own. You don't know when or how it will happen, but you know it will happen. You know it, but you do not understand it. The incomprehensible inevitability of death and the need to know why inspire a different approach to knowledge, called faith. Unlike science, faith does not rely on observation of external reality. Quantification, measurement, and other conceptual tools are irrelevant. Unlike science, faith is ultimately personal, not social. At its core it cannot be shared.

We humans are therefore aware of two categories of ignorance: what we don't know, and what we can't know. We have developed two ways to approach ignorance: science and faith. These two ways lead to two types of knowledge. Science says 'I know because I have tried it and it happens this way.' Faith says 'I know because I believe.' The two types of knowledge are, however, difficult if not impossible to disentangle. At least since Darwin, science has been steadily encroaching into areas of knowledge previously claimed by faith. Yet science always interacts with faith, both individual and collective. 'Culture', for instance, is what you 'know' without thinking about it. Some of what you 'know' you can, if you think about it, verify objectively. On the other hand much of what you 'know' may not be objectively true, by any external criterion, starting with your sense of self. The implications for individuals and for society have preoccupied philosophy and the arts throughout human history.

The interactions of science and faith influence the behaviour not only of individuals but also of groups. Although faith is essentially personal, it is often manifest in ways that elicit a positive response from others, and an impulse to communicate and share some form of mutual understanding based on faith. If enough people can agree on what they consider essential aspects of this mutual understanding, a shared form of faith becomes a religion. For those who adhere to it this religion becomes a key group identity. Because a religion addresses questions that humans feel the need to answer, it can have a potent role as a form of social organization. Throughout history religions have developed bodies of shared knowledge, based on faith, that have shaped human society, its structures and its processes. At its strongest, a religion can be viewed as a parallel form of government, able to influence the behaviour of those under its authority at least as effectively as any secular government, and possibly more so.

Knowledge based on science and knowledge based on faith differ in one crucial respect. Science is a global activity. To be accepted, scientific understanding must apply everywhere. Faith, by contrast, requires no external verification. It need not be shared; but when it is, the essential requirement is mutual belief, not verification. Accordingly, one group of faithful can agree on one body of shared faith or religion, while another group of faithful rejects that body and agrees on some other body. Because of the intensely personal nature of faith, the 'us–them' syndrome may then emerge in a peculiarly virulent form. The perceived disparities of knowledge and understanding may make each group of faithful regard the other as inferior. History abounds with grisly examples of the consequences – brutality, torture and slaughter. Whatever the faith, historical evidence indicates at best mixed results for humanity.

The same may be said of science. Since the Industrial Revolution, science and the technology it fosters have all too often aggravated disparities between 'haves' and 'have-nots'. Science and technology have enabled humans to intervene in natural systems so intensively that some systems now show serious signs of breakdown. Moreover, from edged tools to weapons of mass destruction, scientific knowledge has enabled humans to kill each other more and more easily, and in ever-greater numbers. Unlike faith, science rarely provokes killing; but science facilitates it. Even as science edges relentlessly onto the terrain of human knowledge once dominated by faith, the faithful use science to arm themselves.

Knowledge of whatever kind is manifest in action. How we apply knowledge, and to what purpose, depends on the values that guide our choice and course of action. Sharing a single interconnected planet, no matter what we know or how we know it, we have yet to learn to act wisely.

Values in conflict

Humans act according to what they value – what they consider worth doing, worth having, worthwhile. No matter what they say they value, how they act demonstrates what they really value. The difference in values endorsed by words and values demonstrated by deeds is frequently stark.

If someone asks you seriously what you value, you may answer equally seriously that you value your spouse or mate, your children, your family, your friends – certain particular other humans close to you. Depending on your circumstances, including age and health, you probably value your life. People value their personal property – specific goods of which they can claim ownership. Some individuals value money, status, power or all three. Historically, people have always valued land, especially if they own and control it. In all of these cases, the value you place on someone or something is demonstrated most vividly by how you respond to a threat that endangers either what you value or its relationship to you. If you value someone or something enough, the threat of danger, attack, injury, damage, theft or conquest will prompt you to respond, in an effort to repel the threat or rescue the endangered.

Such values are immediate, specific and personal, and elicit actions accordingly. When the question of values is posed in more general and abstract terms, however, the answers and the actions tend to be much more ambiguous. What values do humans in general recognize? How are they acquired and how manifest? Take an obvious example: laws and customs in most places call murder – killing another human – a crime. This might suggest that humans in general value human life in general. That, however, is clearly an oversimplification. Sometimes, in some circumstances, humans value human life highly; witness for instance innumerable acts of selfless heroism in fires, floods, earthquakes and other disasters. On the other hand, humans are very good at killing other humans, and notably prone to do so. They kill not only complete strangers but those closest to them – wives, husbands, siblings, offspring. Some commentators argue that only social inhibitions and prohibitions prevent humans, especially young males, from resolving every conflict by a straightforward fight to the death with all available weapons.

Military training reverses the social process and removes the inhibitions, making killing socially acceptable in certain circumstances. Perceived disparities also weaken inhibitions. Killing someone you consider inferior is easier. As recent events have demonstrated with pitiless clarity yet again, you do not have to be a trained soldier to kill unarmed civilians.

Even the value of your own life depends on the context. The tenacity with which some individuals cling to life in even the grimmest conditions testifies to the value they place on life. Nevertheless the prevalence of suicide demonstrates that for some individuals, at some time, continued survival is intolerable. Pious pronouncements about the sanctity of human life fail to address the realities of experience. How people value people, and how they demonstrate this valuation, range from unconditional love to implacable hatred, with every imaginable variation in between, and actions to match. Moreover the valuation changes with time. The evidence suggests that the value humans put on human life is at best provisional. It depends on who is doing the valuing, and in what circumstances.

Consider another example. Human societies almost always claim to value children. Children embody the future. They are a source and a focus of pride, and in some contexts also of power. Because they are comparatively vulnerable, they elicit an especially protective reaction from most adults. That said, nevertheless, human societies routinely treat children in ways hard to reconcile with any consistently high valuation. Some societies indeed resort quite ruthlessly to infanticide, especially of female children. Moreover, quite apart from blatant abuse and neglect, even treatment ostensibly well meant may miscarry. Schooling, for example, all too often suffocates a child's hitherto enthusiastic eagerness to learn. For too many children school is frustrating or frightening, a place to learn modes of behaviour from which society will later suffer. If society genuinely valued children it would reappraise and reorganize the process of childhood learning from infancy onwards.

Humans value their own personal property; indeed some humans regard their children as their property, value them on that basis and act accordingly. Etymologically, the word 'property' is akin to the word 'ownership', in the sense of belonging to the self, with the implications that entails. One immediate corollary is that property is by definition exclusive: what's mine is mine, and is therefore not yours. Some commentators suggest that the urge to possess property and especially land, as defensible territory, is part of the genetic make-up of humans, in particular adult males seeking mates. Whatever the truth of this suggestion, humans undoubtedly value possession of land, including the exclusive aspect of possession that demands repulsion of intruders.

Paradoxically, however, this value may be endorsed even if the possession of land is not personal but collective and indirect, through some form of group allegiance, ultimately perhaps a nation or a state. That moves the question of values and their effect on human activity into a different dimension. A single individual can hold certain specific people, things and places valuable by virtue of direct connection to them, in the form of personal relationship or ownership, acting accordingly to cherish and defend what is of personal value to that individual. However, the value manifest through actions inspired by personal relationship or ownership is different in nature to the value manifest through actions inspired by allegiance to a group. Consider, for example, the great abstractions so often acclaimed as variously ultimate values for the whole of humanity – for instance freedom, independence, justice and peace. These values are not personal, but shared within a group. For that reason, actions inspired by these values at once raise issues of group decision-making, leadership, governance and power.

Even the interpretation of the meaning of a value such as 'independence' becomes an issue for the governance of the group. Conversely, disagreement between members of the group about the interpretation and consequent action may affect the leadership and governance of the group. A further complication for such shared values is the frequent gap between the ostensible value that the group through its leadership endorses in words, and the value the group demonstrates by actions, especially when the actions are undertaken by the group as a whole. Disputes between group members about their shared values are usually charged with emotion; evidence of inconsistency or frank hypocrisy, particularly on the part of leaders, intensifies the emotion.

Yet more intense is the emotion arising between different groups that endorse different shared values, or interpret them differently. For humans in groups emotion is contagious, in any context from a football match to a lynching, no matter what the individuals involved may value personally. The power of a charismatic leader arises in significant measure because the leader understands instinctively the system dynamics of a crowd of people, and can manipulate the crowd accordingly. The great abstractions – freedom, independence and the rest – are potent tools for manipulating crowds. Leaders have almost always stressed the disparities between 'us' and 'them', the superior values we share and the inferior values they share, to great and sometimes devastating effect. This approach also serves to distract attention from the genuine disparities, particularly of power and access to resources, within the group itself. An exclusive group identity such as nation or faith, if coupled with perceived disparity in shared values between that group and those it excludes, makes conflict all but inevitable. Most of human history is devoted to recording the consequences.

One crucial corollary, however, is often overlooked. Values such as freedom, independence, justice and peace, shared by people in groups, interact with one another. Their meanings and interpretations change, depending on the context; and so do the actions they inspire. Shared values function together in value systems, which evolve over time. In the rhetoric of leaders these concepts of value, these great abstractions, resound as absolutes. They are not. They are provisional – like all of human life.

Whatever their status as abstract philosophical concepts, values as impetus to action interact and evolve, as value systems. They also interact and evolve with other systems of importance to humans, including social systems, technological systems and natural systems. All these systems are interdependent, including those of unquestionable and essential value to humans both individually and in groups – the natural systems of which we are a part. Human value systems still tend to take natural systems for granted, to ignore or at best to undervalue them. The consequences are alarming.

Unlike social or technological systems, natural systems are inherently global. Yet we do not have a global value system, internally consistent, evolving with the advance of collective knowledge and understanding, and embracing all the interdependencies that make up global society. Instead we have many different value systems, inconsistent and conflicting. We humans, as individuals and groups, act according to these conflicting value systems; and in so doing, we threaten to destroy ourselves. Reconciling these conflicts of value may be the severest challenge we face.

A global agenda

As we enter a new millennium, the defining issue for the global agenda is simple. Do we as humans seek ways to resolve conflicts on a global basis? In other words, do we – at least those of us in a position to do so – accept responsibility for running the planet? In principle, 'we' means all humanity. In practice, 'we' means those of us reading this essay, and those who might. In a world awash in information, knowledge and its application become global issues. Access to information, and the knowledge it can convey, is now itself a source of widening disparity between 'haves' and 'have-nots'. We are the 'haves', and responsible accordingly. We have a better opportunity than most of our fellow humans to be aware of and to understand the challenge, and to attempt to meet it. If, however, those of us among the 'haves' presume some form of superiority because of our disparate access to information or knowledge, we shall not alleviate the disparity but aggravate it, and with it the problems it poses. A measure of humility is essential. We know and understand a lot less than we may realize, and a lot less than we need to know.

We now live in a world of daunting complexity. Natural and human systems interconnect, interact and evolve, carrying us with them at bewildering speed. Whether we like it or not, human society is now unambiguously global; only global catastrophe can break the interconnections. Such catastrophe is all too easy to imagine. Unless we learn to live together, we shall assuredly die together; but dwelling on the possibility is little help. A counsel of despair will be self-fulfilling. If we look for signs of hope, we find them. Although some tensions and stresses continue to mount, others are easing.

In many parts of the world, for instance, the traditional subordination of women to men has begun to abate, gradually reducing the disparity of opportunity between the two halves of humanity. The process is halting and erratic, and by no means universal, but it is happening. Long-standing barriers between different groups of people, including national borders, language differences and cultural divisions are becoming less obstructive and obtrusive. Varieties of the English language are spreading across the globe, offering a common medium of communication. English is becoming a world language for the wrong reasons – because of the military and economic power of Britain in the nineteenth century and the US in the twentieth. But English is endlessly flexible and adaptable, a bridge across the gap of incomprehension that different languages impose. Native anglophones, however, must beware of a disconcerting corollary. If you are a native anglophone and know nothing of the native language of a non-anglophone speaking English to you, the other speaker understands you better than you understand the other speaker. As English becomes a world language, monoglot anglophones, paradoxically, may be at a grave disadvantage.

As barriers fall, the isolation of individuals and groups diminishes and opportunities for cooperation increase, reinforcing interactions and interdependence. Diminishing isolation shifts power structures; it can flatten hierarchies, foster decentralization and enhance personal responsibility and control. Barriers and isolation make offences against humanity easier to commit. If the barriers fall, if the whole world knows, the whole world can react in outrage, and demand that the perpetrators be seized and punished. Such possibilities are as yet at best incipient, but the trend is clear. The early outlines of a global value system may now be perceptible.

Such global convergence should not, however, be misconstrued. What is desirable is not uniformity but equitable diversity. Diversity provides resilience, a robust network that is mutually supportive, reinforcing and enriching. As we strive to reduce disparities we should welcome and celebrate diversity. This is not altruism, or mere tolerance of difference; it is in our own interdependent interest, individually and collectively. Indeed the time may be at hand for a global declaration of interdependence.

If we humans accept the responsibility for running the planet, we must therefore ensure that as many of our fellow humans as possible are successively enlisted and equipped to participate, to contribute and to share the responsibility. We shall need to develop a common global agenda – what needs to be done. Anyone who looks at the world today will have no doubt about the magnitude of the task before us. Nevertheless, humanity can already offer an extraordinary diversity of knowledge, skills and competence. We need to bring it all to bear on the future, communicating and disseminating what we know, and learning what others know, to extend our common understanding. Working together we may even attain a degree of wisdom.

Wisdom, however, always acknowledges uncertainty. Anything you know absolutely is almost certainly wrong. Unsettling though the thought may be, knowledge and values are provisional. Life itself is provisional. Let us take that as a given, and proceed accordingly. We might perhaps begin with two propositions we all can accept. We are all members of the human species; and we are all citizens of the planet Earth. Thereafter the details are open to discussion – from first principles.

No one knows all the answers. We may not even be asking the right questions. We are all in this together, and we'll need all the help we can get.

Annex 2

DISCUSSING ENERGY: A STYLE GUIDE

This style guide offers an annotated glossary of words and phrases used in writing or talking about energy. It indicates the imprecision, inconsistencies and contradictions that bedevil discussion of energy, particularly as a policy issue. Modelled on the style guides used, for instance, by *The Guardian* and other quality newspapers, it is an attempt to eliminate or at least minimize such shortcomings in the language we use to discuss energy. For ease of use, it presents words and phrases in alphabetical order. Terms in bold have separate alphabetical entries.

alternative energy/alternative energy source *(see* **energy source***; 'alternative' means excluding coal, oil and natural gas, and may also exclude hydroelectricity and nuclear energy, depending on the commentator; small hydro and advanced nuclear reactors are commonly classed with the 'alternative sources')*: in common use through the 1990s, thereafter less common; best avoided.

ambient energy *(energy that is present but unnoticed, usually unmeasured and free of charge, for instance the energy of sunlight, wind, warm bodies and other* **energy systems** *warmer than their surroundings)*: some commentators, including this one, prefer this expression to renewable energy, to refer to natural energy flows (not including biomass, which is a **fuel**). The expression 'ambient energy' is not in common use; perhaps it should be.

commercial energy (**energy carrier**, *that is, fuel or electricity, delivered in measured quantities and paid for accordingly)*: energy policy and its precursor, fuel and power policy, have historically focused on measured quantities of fuels and flows of electricity – that is on the sale and purchase of energy carriers by the unit, as 'commercial energy'. 'Energy policy' so defined is really just a shorter way of saying 'fuel and power policy'. Genuine 'energy policy', however, should encompass also the requisite **energy technology**, including infrastructure, not only to produce and deliver **energy carriers** but also to provide the **energy services** that users actually want. Technology and infrastructure are also 'commercial',

This annex is adapted from first publication in *The Energy Alternative: Changing the Way the World Works* (Boxtree, 1990, Optima, 1991).

but their financial structure centres on investment in and management of assets, not on commodity transactions by the unit.

conversion *(what actually happens in energy processes – energy is not 'consumed', but converted from one form to another)*: the expression 'energy conversion' is not in common use, but perhaps should be, when referring to 'energy' – for instance **ambient** or **renewable energy** – rather than **fuel**.

delivered energy *(amount of* **energy carrier** *reaching customer's meter, weighbridge or other measuring device; amount of energy carrier for which customer must pay supplier)*: note that **ambient energy** is delivered unmeasured and free, but may be an important contributor to energy services such as comfort and illumination. 'Delivered energy' is just a shorter way of saying 'delivered fuel and electricity'; it does not include all relevant energy flows.

efficient/efficiency *(desirable attribute of* **energy system** *in which final* **energy service** *is provided by a combination of* **energy technology**, **ambient energy** *and* **energy carriers** *to achieve an optimum according to some physical or economic criterion; may also, however, be used as all-purpose hooray-words in policy pronouncements, and become roughly equivalent to NEW IMPROVED as used by detergent manufacturers)*: should be used sparingly and precisely, where the 'efficient' process is clearly identified. 'Efficiency' is strictly speaking a ratio, less than 1, comparing the useful output of a process to the total input. Any context in which no such ratio can be clearly identified – for instance the performance of a building – is a dubious context in which to use 'efficient' or 'efficiency'. Use 'performance' – qualitative, not quantitative – instead.

energy *(oil;* **commercial fuels**; *electricity;* **commercial energy carriers** *of all kinds; all* **energy carriers**, **commercial** *or otherwise;* **ambient energy**; *ambient* **energy technology** *(solar panels, wind generators and so on); intermediate* **energy technology** *(refineries, power stations and so on); combinations and permutations of the above; almost never 'energy' in its strict physical scientific sense)*: as the foregoing indicates, 'energy' has become an all-purpose word whose meaning in any given context may differ widely from that in any other context, even within the scope of energy policy itself, to say nothing of everyday speech. It is also a convenient shorthand for various subcategories as above. This convenience, however, comes at the cost of serious imprecision in what is being discussed. Moreover, if 'energy' is understood to mean only measured flows of measurable energy, it may misrepresent the overall function of an energy system. It may also obscure opportunities for improvement, and arbitrarily restrict the scope of energy policy.

energy carrier *(a material or phenomenon that can store energy or transport it from place to place, usually by implication under human control; all fuels plus electricity)*: strictly speaking, to be sure, sunlight, wind and other natural phenomena are also energy carriers, and so indeed is the blood stream; but the expression – which might perhaps be more widely used – is likely to be most useful if applied specifically to carriers under human control.

energy conservation *(***'energy***' in its strict physical sense is invariably conserved in any energy* **conversion** *process, whether or not under human control; as used in policy pronouncements 'energy conservation' usually means 'energy carrier conservation', that is, using less fuel or electricity, as a result of various measures, negative or positive, short term or long term; all too often merely hooray-word, otherwise undefined)*: best regarded warily and used sparingly if at all.

energy consumption *(amount of fuel used, directly or indirectly; amount of fuel plus electricity used; amount of electricity used; amount of* **commercial energy carriers** *used; amount of* **energy carriers***, commercial and non-commercial, used; literal meaning is scientifically wrong)*: avoid. 'Energy use' is preferable.

energy demand *(in the past, recorded purchases of fuel or fuel and electricity; in the future, anticipated purchases of fuel or fuel and electricity;* **energy carriers** *converted by final users; can be specified before intermediate* **conversion***, for instance power station, or after, leading to very different results; note that in the future unsatisfied 'demand' cannot exist since demand is identified and quantified only by satisfying it)*: avoid. If necessary, either 'energy use' in the past or 'anticipated energy use' in the future more accurately describes the concept.

energy production *(extraction and processing of fuel; intermediate* **conversion** *of fuel into a* **secondary energy carrier** *such as electricity; controlled* **conversion** *of* **ambient energy***, for example generation of hydroelectricity or wind electricity; combinations and permutations of the above; literal meaning is scientifically wrong)*: 'production' is defensibly correct when referring to fuel or electricity, but not when referring to 'energy'. 'Energy production' is a shorthand that smears together quite different processes, usually for the purpose of aggregated statistics whose meaning then becomes ambiguous and misleading for policy purposes – not least because it suggests substitutability where none may exist. For accuracy, use 'fuel production' and 'electricity generation'. See also **energy supply**.

energy producer: strictly, one who contravenes the first law of thermodynamics. Use 'fuel producer' or 'electricity generator' as appropriate.

energy services *(what energy users actually want – comfort, cooked food, illumination, motive power, mobility, information processing and so on)*: sometimes used to refer to the supply of **delivered energy**, particularly in the form of gas or electricity, but should not be so used.

energy source *(fuel; electricity – a questionable usage at best, and in the case of electricity generated from fuel scientifically wrong; active* **ambient energy** *technology, for instance solar panel – but not usually the ambient energy itself, and almost never the sun, although* *the sun is the source of almost all the energy converted on Earth)*: too prone to misconstruction; avoid.

energy supplier *(supplier of* **commercial** *fuel or electricity)*: better to use 'fuel supplier', 'electricity supplier', or 'energy service provider' as appropriate.

energy supply *(fuel available for use; fuel plus electricity available for use;* **commercial energy carriers** *available for use; all* **energy carriers** *available for use; may also include* **ambient energy** *consciously converted for use; does not – repeat not – embrace the* **energy conversion** *which makes the Earth habitable and constitutes more than 99 per cent of the total energy conversion taking place on it)*: a shorthand that smears together quite different processes, usually for the purpose of aggregated statistics whose meaning then becomes ambiguous and misleading for policy purposes – not least because it suggests substitutability where none may exist. 'Fuel supply' is correct when applicable, as is 'electricity supply'; but they are not interchangeable. See also **energy production**.

energy system: concept that should be at the heart of real 'energy policy', as distinct from 'fuel and power policy'. An energy system includes all **ambient energy**, **energy technology** and **energy carriers** involved in an energy process; energy policy should recognize explicitly and address all the relevant factors, as a coherent and integrated whole, rather than focusing only on energy carriers while ignoring the ambient energy and energy technology that are essential features of the complete system.

energy technology *(structures, devices or systems designed, fabricated and operated by humans to intervene in and control the conversion of energy for human purposes)*: usually used to refer to technology involving commercial energy carriers; but usage should be expanded to include also human artefacts that manipulate ambient energy without measuring it – for instance buildings.

fuel *(etymologically, 'material for a fireplace'; material whose energy content can be mobilized where and when it is desired for use; note, however, that some commentators use 'fuel' to mean only* **commercial** *fuel, or to include also electricity, which is not a material and cannot in practice be stored)*: a precise and specific term; use instead of 'energy' when applicable. 'Fuel' is not, however, a synonym for **energy carrier**; in particular, do not use to include electricity.

non-renewable energy/non-renewable energy source *(fossil fuel; not usually applied to deforestation for firewood, or siltation of hydroelectric installations; sometimes applied to nuclear energy generated by using uranium in conventional nuclear reactors)*: no longer in common use; best used only in explicit contrast with **renewable energy**, and even then sparingly and with care, because of possible ambiguity and imprecision.

power *(grid electricity; all electricity, grid or otherwise; used all too frequently as equivalent to* **energy***; occasionally, albeit rarely, used with physical scientific meaning, that is, rate of* **energy conversion** *– energy converted per unit of time, for instance joules per second or watts)*: in energy policy, 'power' usually refers to traditional centrally generated electricity delivered to users over a network, as in 'the power sector'. In general, use 'electricity' rather than 'power'.

primary energy *(virgin fuel; hydroelectricity or nuclear electricity sent out from generating stations; sometimes stated net of energy used in production or generation, sometimes not; sometimes includes non-commercial fuel, sometimes not)*: a synthetic term, used for statistical purposes to aggregate forms of energy whose only common attribute is that they are measured, usually in a commercial context. Suggests substitutability that may not exist. In common use, but best regarded warily.

renewable energy/renewable energy source *(sunlight and its derivatives, sometimes embracing the relevant technology, sometimes not; the ambiguity is unfortunate, since so-called 'renewables' other than biomass are based on the conversion of* **ambient energy** *that is itself free; the costs arise in controlling its conversion for use, for example by wind generators or solar cells)*: a generic term in increasingly common use; but see **ambient energy**.

secondary energy *(fuel or electricity produced by intermediate converter such as refinery or power station; sometimes stated net of losses in processing and delivery, sometimes not)*: used for statistical purposes to aggregate forms of energy whose only common attribute is that they are measured, usually in a commercial context; see **primary energy**. Suggests substitutability that may not exist. Best regarded warily.

useful energy *(energy whose conversion identifiably furthers user's objective; delivered energy less energy lost by end-use technology – up chimney, in friction and such)*: when quantified, used for statistical purposes; but most useful energy is ambient energy – not measured, not quantified, not paid for.

Annex 3

GLOSSARY

Alternating Current (AC): electric current that surges back and forth, usually at either 50 or 60 times a second, the **frequency**, called 50 or 60 hertz and written 50Hz or 60Hz.

Alternator: coils of heavy wire carried on a spinning shaft between magnets. As the shaft spins, at 50 or 60 revolutions per second, electric current surges back and forth in the coils and the wires connected to them, once in each direction per revolution – alternating current or AC. The shaft can be spun by a water turbine, steam turbine, gas turbine, diesel engine or wind turbine.

Amperage: quantity of electric current flowing.

Ampere or **Amp**: unit of electric current, written A.

Ancillary Services: on a **synchronized AC** electricity system, various functions that have to be provided by **generators** to maintain system stability, including **voltage** control, **frequency** control and a number of others.

Base Load: the minimum continuous **load** on an operating electricity system.

Biomass Power: uses biomass fuel – wood or other vegetable matter, including wood residue. The fuel can be simply burned in a combustor to raise steam for a steam turbine coupled to an alternator; or it can be 'gasified' in a 'gasifier' – see **integrated gasification combined cycles** – to produce combustible gas, to be burned in a gas turbine or a diesel engine coupled to an alternator.

Black Start: the ability of a power station to start up from a complete shutdown in isolation, without using electricity from the network – essential to recover from a system shutdown.

Cable: one or more strands of wire that conduct electricity, often covered with one or more layers of insulating material.

Capacity Factor: see **load factor**.

CCGT (combined-cycle gas turbine): see **combined cycles**.

This annex is adapted from first publication in *Transforming Electricity* (RIIA/Earthscan 1999).

CHP: **combined heat and power**; **cogeneration**.

Circuit: a loop around which electric **current** flows.

Cogeneration: generation of electricity plus usable heat, in the form of steam or hot water, from the same quantity of fuel in a single operation.

Combined Cycles: combining one or more gas turbines with one or more steam turbines. The exhaust gas from the gas turbine is still at a temperature of over 500 Celsius, and can be passed through a heat exchanger to raise steam for a steam turbine, from the same fuel. Combined cycles can raise overall fuel efficiency to perhaps 60 per cent.

Combined Heat and Power: see **cogeneration**.

Current: flow of electricity.

Diesel Engine: an internal combustion engine like a car engine, but much larger, burning diesel fuel.

Direct Current (DC): electric current that flows always in the same direction.

Dispatch: instruct a **generator** to begin delivering electricity into the system, or to stop doing so.

Distribution: delivery of electricity to users at low **voltage**.

Distributed Generation: smaller-scale production of electricity, usually much less than 50MW per unit, usually connected to a network at less than **transmission voltage**.

District Heating: delivery of steam or hot water through a network of pipes to heat a number of buildings in a district.

FACTS: flexible **AC transmission** system, using **power electronics** to enable subtle control of current flows through system.

Frequency: of **alternating current**, the number of times the current surges back and forth in a second.

Fluidized-Bed Combustor (FBC): a combustion chamber with a layer or 'bed' of inert granular material such as sand or ash in the bottom. Air blown up from below lifts the bed material into a churning mass; fuel burned above or in the bed makes the bed material incandescent. When low-grade fuel, such as poor-quality coal or combustible residue, is fed into the FBC, the incandescent bed material ignites it immediately; even wet fuel cannot quench the combustion. Limestone or other calcium-bearing mineral added to the bed captures sulphur as solid calcium sulphide, minimizing sulphur dioxide emissions. The combustor is lined with water tubes, and the hot combustion gases pass over other heat exchangers, to boil water and raise steam for a steam turbine.

Fuel Cell: a device in which a flow of hydrogen gas reacts continuously with oxygen from the air, not in a flame but in a so-called 'electrochemical reaction' through a porous material called an 'electrolyte'. The reaction creates an electrical pressure or voltage between the two outer terminals of the cell; the voltage drives an electric current through a circuit connecting the terminals. Generating technologies that use mechanical power to turn a shaft coupled to a rotating alternator produce alternating current or AC; a fuel cell, however, produces 'direct current' or DC. Different types of fuel cell are now under development, described according to the different electrolytes they use, including 'molten carbonate', 'solid oxide' and 'polymer'. Each has advantages and disadvantages for different applications. Because hydrogen gas is not at the moment readily available, a fuel cell usually requires a 'reformer' that can make hydrogen from another fuel, such as petrol or natural gas. The DC output of the fuel cell can be used directly, to power equipment that requires DC, such as electronics; alternatively, the fuel cell can be connected to the AC network through an **inverter** that turns the DC into AC. The only emission from a hydrogen-powered fuel cell itself is water vapour. Emissions from the reformer depend on its design and the fuel it uses, but can be minimal.

Gas Engine: an internal combustion engine like a car engine, but much larger, burning natural gas.

Gas Turbine: a shaft carrying two sets of blades inside a close-fitting casing. The first set of blades, the 'compressor', is a series of windmills of decreasing size; air entering from the atmosphere passes through the compressor section, which compresses it to perhaps twenty times atmospheric pressure. This compressed air passes into a combustion chamber, where fuel is mixed with it and burned. The hot high-pressure combustion gas then expands through a second set of complex blades, the 'gas turbine' blades, able to withstand temperatures of 1200 Celsius or higher, spinning the entire shaft, including the compressor and an alternator also coupled or attached to the shaft. The gas turbine can burn light liquid fuel or gaseous fuel, including natural gas and gas from a 'gasifier' – see **integrated gasification combined cycles**.

Generator: technology that produces electricity.

Gigawatt: a billion **watts**, written GW.

Green Electricity: refers to electricity produced by **generators** considered to have acceptably low environmental impact; the practical definition is controversial.

Grid: **network** of high-voltage **transmission** lines.

Hertz: unit of **frequency** of **alternating current**, written Hz.

High-Voltage Direct Current (HDVC): **direct current** at a **voltage** high enough for **transmission** without unacceptable losses, usually more than 100,000 volts.

Hydroelectricity, **Hydro**: electricity generated by water turbines.

Hz: **hertz**, unit of frequency.

Independent Power Producer: a **generator** delivering electricity to a system but not owned by the system.

Infrastructure Electricity: not yet in common use; refers to electricity produced not by converting fuel but by converting ambient energy in suitable technology.

Integrated Gasification Combined Cycles (IGCC): a combined cycle arrangement in which the fuel for the gas turbine is a combustible gas produced in a 'gasifier'. In a gasifier, a comparatively awkward fuel such as coal, refinery residue or biomass is partially burned with limited oxygen, to produce a combustible mixture of carbon monoxide and hydrogen, that is fed into the combustor of the gas turbine. An IGCC unit can capture sulphur from the combustion gas using conventional refinery technology.

Intelligent Meter: **meter** able not only to measure the cumulative flow of electricity, but to gather other information about the services provided, and possibly also to control them as desired.

IPP: **Independent Power Producer**.

Inverter: unit that converts **direct current** to **alternating current**.

Kilowatt: unit of **power**, 1000 **watts**, written kW.

Kilowatt-Hour: unit of electrical energy, written kWh.

Lamp: converts electrical energy to light energy.

Load: technology that uses electricity; also, the amount of electricity it uses when connected to the system.

Load Curve: pattern of changing total **load** on the system, hour by hour, day by day and month by month.

Load Factor: actual amount of electricity produced, compared to maximum amount possible over the same period, expressed as a percentage; when applied to a single generator, often called **capacity factor**.

Load-Following: changing the output of a **generator** to match the changing **load**.

Megawatt: a million **watts**, written MW.

Megawatt-Hour: one thousand **kilowatt-hours**, written MWh.

Merit Order: in traditional electricity, the order in which power stations are **dispatched**, as **load** changes from **base load** to **peak load** and back.

Meter: device to measure the cumulative flow of electricity.

Microcogeneration: uses a small internal combustion or Stirling engine or microturbine to produce both electricity and usable heat or hot water from the same fuel.

Microturbine: a small 'gas turbine' with one moving part, a single shaft rotating at very high speed, whose high-frequency AC output is then converted electronically into conventional AC.

Motor: converts electrical energy to mechanical energy of a rotating shaft.

Network: cables and wires that link **generators** and **loads**.

Nuclear Power: electricity generated by using heat from a nuclear reactor to produce steam for a steam turbine.

Outage: shutdown, scheduled or unscheduled, of a **generator**.

Peak Load: the maximum load on the electricity system, the most electricity the system has to deliver at a given moment in time.

Photovoltaics (PV): uses light, usually sunlight, falling on a thin sheet or 'cell' of specially prepared material, a so-called 'solar cell' or 'PV cell', to produce an electrical pressure or voltage between connectors attached to different layers of the cell. A 'solar panel' or 'PV panel' consists of dozens or hundreds of individual cells, interconnected into a single electrical circuit – the more cells, the higher the output of electricity, for a given brightness of light. Like a fuel cell (see above) a PV array produces DC electricity. PV cells are now being integrated into components of buildings, in the form of PV tiles for roofs and facades. PV tiles perform a dual function; they not only form part of the building structure itself, but can generate a significant fraction of the electricity required for equipment inside the building.

Pool: traditionally, a cooperative arrangement whereby monopoly-franchise electricity systems exchange electricity for backup and load-levelling; on a liberalized system, an arrangement whereby competing generators bid to establish which are to be connected to the system at which times.

Power: the amount of energy delivered in a unit of time, measured in **watts**, written W.

Power Electronics: switching devices like computer chips, but able to handle heavy electric currents.

Power Pack: converts **AC** electricity from the system into low-voltage **DC** electricity for electronic appliances.

Power Quality: an indication of the amount of disturbance carried on a **synchronized AC** system; high quality means minimal disturbance.

Power Station, Power Plant: one or more large **generators**, connected through a **switchyard** to an electricity system.

Profile: in electricity, refers to some form of averaged **load curve** on a user's premises, usually for purposes of billing.

Rectifier: device that converts **AC** electricity to **DC** electricity.

Renewable: refers to energy that is replenished continuously by natural systems; renewable energy technologies recover this energy either directly as electricity (**hydroelectric power, wind power, solar thermal power, photovoltaics**) or as fuel (**biomass power**).

Resistance: opposition to the flow of an **electric current**.

Solar Thermal Power: uses mirrors or other devices to concentrate sunlight, to boil water into steam or other working fluid into hot pressurized gas, to turn a turbine coupled to an alternator.

Spinning Reserve: a large **generator** whose shaft is turning, **synchronized** with the electricity system, but delivering no **power**, held in readiness in case another **generator** fails.

Stability: the condition of an electricity system in normal operation, when voltages, frequencies and other important operating attributes are at or very close to their desired values throughout the system.

Steam Turbine: a shaft carrying a series of blades like windmills of increasing diameter inside a close-fitting casing, spun by steam expanding through the series of windmills. A 'condensing turbine' has cooling pipework at its outer end; when the expanding steam reaches the cooling pipes the steam condenses to water, lowering the pressure dramatically, so that steam still coming through the turbine meets no opposing pressure. A 'back-pressure turbine' allows the steam to leave the turbine without condensing, in order to use the steam elsewhere, for industrial process heat or district heating. A 'pass-out turbine' has one or more ducts part-way along it, from which steam can be tapped at intermediate pressure, usually for industrial process heat. The steam is produced in a 'boiler', in which heat boils water and superheats the steam as required. The heat can come from burning fuel – usually coal, oil, natural gas or biomass – or from a nuclear reaction in uranium, in a 'nuclear reactor'.

Stirling Engine: an 'external combustion' engine, in which a heat source outside a cylinder heats a working fluid inside the cylinder to drive a piston back and forth; in a Stirling engine generator the moving piston in turn operates a generator.

Switch: a junction in an electrical network that can be closed to allow the flow of current through the junction, or opened to interrupt it.

Switchgear: equipment including heavy-duty **switches** able to allow or interrupt large flows of current without damage.

Switchyard: at a **power station**, **switchgear**, **transformers** and **cables** to control and direct the flow of electric current from the power station into the rest of the electricity system, altering its voltage as necessary.

Substation: an installation on an electricity system but not at a **power station**, with **switchgear**, **transformers** and **cables** to control and direct the flow of electric **current** at that point on the system, and alter its **voltage** as desired.

Synchronized: surging back and forth exactly in step.

System Services: see **Ancillary Services**.

Transformer: a device that can raise or lower the **voltage** of **AC** electricity, with a corresponding inverse effect on the current.

Transmission: transport of electric **current** at **high voltage**.

Transmission Line: **cables** usually carried on tall towers to transport electric **current** at high **voltage**.

Trigeneration: production of electricity, usable heat, and usable cold from the same quantity of fuel in a single operation.

Utility: often used to refer to electricity, as well as to gas, water, drainage, public transport and other activities for the 'public good', but defined so vaguely as to be little use.

Voltage: electrical pressure between different points in a **circuit**, measured in **volts**, written V.

Volt: unit of electrical pressure, written V.

Water Turbine: a shaft carrying blades like a ship's propellor inside a casing, spun by water either falling from behind a dam or flowing along a river.

Watt: unit of **power**, written W.

Wind Turbine: a shaft carrying blades turned by the wind – sometimes called a 'windmill', but the shaft turns an alternator, not a mill; now available in sizes ranging from about 1 kilowatt up to 3 or more megawatts.

Annex 4

FURTHER INFORMATION

In only a decade or so, the concept of decentralized energy, electricity and heat has burgeoned all over the world. In the mid-1990s you would have found few if any references or citations even mentioning the expression. Now if you google 'decentralized energy' you almost instantly find well over a million references on the internet alone. With such a wealth of information now readily available, selecting a handful to mention here is of necessity both personal and arbitrary. What follows is therefore one person's entry-level catalogue of resources for anyone wanting to pursue further the ideas presented between these covers. Almost any of the sources mentioned here will point you to others, in a rapidly expanding universe of both general and specific information, commentary and debate.

My book *Transforming Electricity* (Earthscan/RIIA, 1999), which set the stage for *Keeping The Lights On*, included an extensive selection of further reading, classic titles and personal favourites, revisited and updated here. The literature of electricity is vast, and written primarily for specialists – technologists, operators, regulators and so on. Most of the titles mentioned here, however, should be accessible, at least in general terms, to anyone who has read *Transforming Electricity* and *Keeping The Lights On*. This selection makes no pretence to be comprehensive. Many of these titles also include references and bibliographies identifying other titles of interest. See also our website archive Walt Patterson On Energy, www.waltpatterson.org, and its list of links.

No single history of world electricity appears to exist. *Thomas A. Edison: A Streak of Luck*, by Robert Conot (Da Capo, 1979), is one of several biographies of Edison, balanced, vivid and readable, describing his shortcomings as well as his undoubted genius. *The Man Who Invented the Twentieth Century: Nikola Tesla*, by Robert Lomas (Headline, 1999), is a recent biography of the remarkable Tesla, progenitor of AC and many other advances. *Networks of Power*, by Thomas P. Hughes (Johns Hopkins University Press, 1983), is a comparative historical account of the spread of electricity systems in the US, the UK and Germany to 1930, packed with detail and copiously illustrated, let down only by occasional digressions into historical theory that do not contribute much to the remarkable story being narrated. *Electrifying America*, by David E. Nye (MIT

This annex is adapted from first publication in *Transforming Electricity* (RIIA/Earthscan 1999).

Press, 1990), recounts the emergence of electricity in the US, and discusses its social and cultural context – an informative and entertaining read. *The Electric City*, by Harold L. Platt (University of Chicago Press, 1991), is a detailed and massively annotated account of electricity in Chicago, of Commonwealth Edison and its hugely influential, legendary boss Samuel Insull. A more recent historical overview of electricity in the US is *From Edison to Enron*, by Richard Munson (Praeger, 2005), from an advocacy position attacking monopolies and favouring further change.

Electricity Before Nationalization, by Leslie Hannah (Macmillan, 1979), describes the decades of in-fighting over electricity in the UK to 1947. Its sister volume *Engineers, Managers and Politicians* (Macmillan, 1982) continues the story into the early 1960s, with an epilogue through the 1970s – a critical history that does not pull punches. *The British Electricity Experiment*, by John Surrey et al (Earthscan, 1996), is the definitive commentary on the first five years of the headlong liberalization of electricity in the UK. *The Electric Century*, by John Negru (Canadian Electricity Association, 1990), is a popular account of electricity in Canada, again copiously illustrated, published to celebrate the centenary of the Canadian Electricity Association. Less celebratory is *Hydro: The Decline and Fall of Ontario's Electric Empire*, by Jamie Swift and Keith Stewart (Between The Lines, 2004), recounting the disintegration of one of the largest electricity systems in the world. Historical accounts of electricity in other individual countries, and of individual companies, abound; see any good reference library, or the internet.

Our Common Future, the report of the World Commission on Environment and Development, chaired by Gro Harlem Brundtland of Norway, published in 1987 by a number of publishers around the world, first brought the concept of 'sustainable development' to popular attention. *Energy for a Sustainable World*, by José Goldemberg of Brazil, Thomas B. Johansson of Sweden, Amulya K. N. Reddy of India and Robert H. Williams of the US (Wiley Eastern, 1988), was a sweeping, prescient and visionary overview of the emerging relationship between sustainability and energy, and a landmark in the global policy debate. A decade later the same team, with many other eminent colleagues, prepared the *World Energy Assessment* (UNDP and World Energy Council, 2001), a magisterial study, subtitled 'Energy and the challenge of sustainability', global, comprehensive and readable. *Energy for Sustainable Development*, edited by Johansson and Goldemberg (UNDP et al, 2002), offers a concise 'policy action agenda', to which I also contributed.

Long a deeply conservative influence on global energy policy, the World Energy Council at last emerged as an impressively forward-looking body with its report *Energy for Tomorrow's World* (Kogan Page, 1993). Its subsequent reports on *Renewable Energy Resources: Opportunities and Constraints 1990–2020* (WEC, 1993) and *Global Energy Perspectives to 2050 and Beyond*, a joint study with the International Institute for Applied Systems Analysis (WEC/IIASA, 1995), reinforced its reputation. In 2001, as well as the *World Energy Assessment* (see above), WEC published its own study *Living in One World: Sustainability from an Energy Perspective.*

Those tracking the evolution of thinking about electricity should know about *Electricity: Efficient End-use and New Generation Technologies and Their Planning Implications*, edited by Thomas B. Johansson, Birgit Bodlund and Robert H. Williams (Lund University Press, 1989). It was a ground-breaking reappraisal of the state of the art of technologies for electricity use and generation, and the longer-term implications of their emergence. The paper it included on 'The coming reformation of the electric utility industry' by Måns Lønnroth prefigured unambiguously the upheaval now raging worldwide. *Renewable Energy for Fuels and Electricity*, by Thomas B. Johansson et al (Earthscan/Island Press, 1993), a 1160-page encyclopaedia prepared by many of the world's foremost experts at the request of the United Nations, instantly became the bible of renewable energy advocates everywhere – a treasure trove of information and analysis. The data are long since out of date, but the commentary remains a benchmark, even in a field as fast-moving as renewable energy. *Renewable Energy*, by Bent Sørensen (Elsevier, 3rd edition 2004), is more up to date and equally authoritative, particularly for professionals in the field. Many titles discuss specific types of renewable energy; Google will give you more than enough.

Factor Four, by Ernst von Weizsäcker, Amory B. Lovins and L. Hunter Lovins (Earthscan, 1997), is an enthralling tour de force, demonstrating with a cornucopia of striking specific examples how we can double our global wealth while using only half as many resources. Generating and using electricity both figure prominently. After the follow-up, *Natural Capitalism*, with Paul Hawken (Earthscan 1999), Lovins and his team at the Rocky Mountain Institute then focused specifically on electricity, producing the landmark report *Small is Profitable* (RMI, 2002), the definitive analysis of what its subtitle calls 'the hidden economic benefits of making electrical resources the right size'.

For a witty, outspoken and entertaining account of recent developments in energy and electricity, try *Power to the People*, by Vijay Vaitheeswaran of the *Economist* (Earthscan, 2005).

The International Energy Agency, www.iea.org, offers a vast library of reports and studies on energy issues including electricity. They are listed by topic at www.iea.org/Textbase/subjectqueries/index.asp. The World Bank, the UN Development Programme and the UN Environment Programme produce a variety of reports relevant to global electricity issues, as do international environmental organizations such as Friends of the Earth and Greenpeace; google them for up-to-date information about their material.

Periodical publications tend to be written for specialists, and published at specialist prices. One that is widely available and affordable, however, is the *Financial Times* newspaper; almost every day it carries reports or commentary on the latest developments in electricity in the UK, Europe and farther afield. Wilmington Media publishes the glossy monthly *Modern Power Systems*, www.modernpowersystems.com. PennWell publishes *Power Engineering*

International, www.peimagazine.com, *Cogeneration and On-Site Power Production*, www.cospp.com, and *Renewable Energy World*, www.renewable-energy-world.com, offered free to those in the business. *Power* magazine www.powermag.com is US-based but global. Platts, www.platts.com, publishes a family of cutting-edge newsletters on energy, including electricity. In the UK the Energy Institute, www.energyinst.org.uk, publishes the monthly *Energy World*. In the US the Electric Power Research Institute, www.epri.com, is a vast source of expert information; see for instance the section on 'distributed energy resources', www.epri.com/der-ppp/index.html. At Electricity Online, www.electricity-online.com, Elsevier publish the journals *Energy Policy, Utilities Policy* and *Electricity Journal*.

The Environmental Change Institute of Oxford University has published a landmark study showing how the UK could cut the carbon dioxide emissions from households by 60 per cent by 2050, with straightforward policy measures, feasible and economic. You can find it at *40% House*, www.40percent.org.uk.

Wind Power in Power Systems, edited by Thomas Ackermann (Wiley, 2005), is a massive compendium that does just what it says in the title. A more improbable title is *Sustainable Fossil Fuels*, by Marc Jaccard (Cambridge University Press, 2005), arguing persuasively for the major technical fixes that would disconnect fossil fuels from climate change.

On decentralized energy, you can find many publications available for free download, such as *A Microgeneration Manifesto*, www.green-alliance.org.uk/grea_p.aspx?id=350; *Small or Atomic: Comparing the Finances of Nuclear and Micro-generated Energy*, www.green-alliance.org.uk/grea_p.aspx?id=338; *Decentralizing UK Energy*, www.greenpeace.org.uk/climate/solution/index.cfm; and *Unlocking The Power House*, www.sussex.ac.uk/spru/1-4-7-1-10-2.html.

The incisive report on the economics of climate change, led by Sir Nicholas Stern, former chief economist of the World Bank, can be downloaded from www.hmtreasury.gov.uk/independent_reviews/stern_review_economics_climate_change/stern_review_report.cfm. For any long URL such as these above, try just googling the title, to save fussy typing and errors.

Events move rapidly in global electricity, and the internet is your best starting point. Google will now point you to more information than anyone can possibly keep track of. To get you started, here are a few more places to look:

Chatham House	www.chathamhouse.org.uk
European Commission Energy Directorate	http://ec.europa.eu/energy/index_en.html
International Institute for Applied Systems Analysis	www.iiasa.ac.at
International Institute for Sustainable Development	www.iisd.org

Office of Gas and Electricity Markets (UK)	www.ofgem.gov.uk
Rocky Mountain Institute	www.rmi.org
World Alliance for Decentralized Energy	www.localpower.org
World Energy Council	www.worldenergy.org

Some of these sites also offer lists of links. See Walt Patterson On Energy, www.waltpatterson.org, under 'Links' for more.

Good hunting!

INDEX